Approaches of Adaptive Modeling for Sustainable Environmental Systems

This book provides a comprehensive overview of adaptive modeling techniques and explores various modeling approaches, including system dynamics, agent-based modeling, and integrated assessment modeling, with a focus on key environmental challenges.

It examines how adaptive modeling can inform evidence-based policy-making, foster stakeholder engagement, support adaptive governance structures, and facilitate the implementation of sustainable practices. The book presents real-world case studies and examples from diverse environmental contexts to illustrate the practical application of adaptive modeling approaches.

Key Features:

1. Provides an overview of adaptive modeling approaches for sustainable development
2. Emphasizes policy and governance integration for adaptive modeling to achieve sustainable outcomes
3. Proposes the development of sustainable and resilient environmental systems
4. Explains how adaptive modeling informs policy, fosters participation, and supports adaptive governance structures
5. Includes both theoretical and practical case studies

This book is aimed at graduate students and researchers in environmental and civil engineering.

Current Directions in Climate Change Research
Series Editor: Akhilesh Kumar Yadav and Satya Prakash Maurya

This book series intend to address the problems covered under SDG-13 and interconnecting SDG-6, 9 & 11 for effective actions/decision making which is based on latest spatio-temporal analysis and simulation modeling for a wider set of societal problems that play a vital role to achieve sustainable development goal against climate change impacts. It covers modern computational technique and data modeling for analysis and simulation pragmatic solution in spatial domain for a wider set of problems has been emerged more relevant for engineers and scientists.

Key Features:

- Covers recent Approaches in Dynamic Adaptive Modeling in Climate Change via a Machine Learning perspective;
- Includes urban Sustainability in Changing Landscape under Climate Change: Institutional Policies, Workflows and Implementation using Artificial Intelligence;
- Describes Natural Resource Scarcity, Contamination and Management; and
- Develops Future Frameworks to Combat Climate Change Impacts

Approaches of Adaptive Modeling for Sustainable Environmental Systems Policy and Governance
Edited by Satya Prakash Maurya, Akhilesh Kumar Yadav, Ramesh Singh, Prashant Kumar

Approaches of Adaptive Modeling for Sustainable Environmental Systems

Policy and Governance

Edited by

Satya Prakash Maurya,
Akhilesh Kumar Yadav, Ramesh Singh,
and Prashant Kumar

CRC Press
Taylor & Francis Group
Boca Raton London New York

CRC Press is an imprint of the
Taylor & Francis Group, an **informa** business

First edition published 2027
by CRC Press
2385 NW Executive Center Drive, Suite 320, Boca Raton FL 33431

and by CRC Press
4 Park Square, Milton Park, Abingdon, Oxon, OX14 4RN

CRC Press is an imprint of Taylor & Francis Group, LLC

ISBN: 9781032866475 (hbk)
ISBN: 9781041384717 (pbk)
ISBN: 9781042013234 (ebk)

DOI: 10.1201/9781042013234

Typeset in Palatino
by Newgen Publishing UK

Dedicated

to

Environmental Engineers and Scientists

around the World

Contents

About the Editors

Satya Prakash Maurya earned his B. Tech. in Computer Science & Engineering, M. Tech. in GIS & RS. He completed his PhD in Geoinformatics under the research domain of Spatial Decision Support System from Indian Institute of Technology (BHU), Varanasi (India). He has also studied Civil Engineering from IEI (India), Kolkata. He has worked as an Assistant Professor in CSED at PCE, Jaipur (India) with research interest in computational approach in machine learning, data modeling, expert systems, and artificial intelligence with application of urban water modeling and futuristic prediction and soft computing. Presently, he is working as a Project Scientist-I in NSDI funded project at IIT (BHU). He has published a number of research papers and is also a reviewer in a few reputed journals. He is an associate member in professional bodies such as International Association of Engineers, the IRED, USA and IEI, India.

Akhilesh Kumar Yadav graduated with a Bachelor of Technology in Electronics and Communication Engineering from the Chaudhary Charan Singh University, Meerut, India. He subsequently obtained a Master of Technology in Environmental Engineering from Madan Mohan Malaviya Engineering College, Gorakhpur, India (affiliated with Dr. A. P. J. Abdul Kalam Technical University, Lucknow). His doctoral degree, with a research focus on air pollution, was awarded by the Indian Institute of Technology (Banaras Hindu University), Varanasi, India. Dr. Yadav has served at various institutions, including Indian Institute of Technology (Banaras Hindu University), Varanasi, India; Indian Institute of Technology Bombay, Mumbai, India; Bhabha Atomic Research Centre, Mumbai, India (as a research collaborator); and Chaoyang University of Technology, Taichung, Taiwan. He has authored/co-authored several research and review articles, book chapters, and edited/authored books, and holds a patent. Dr. Yadav received the Young Engineer Award from the Institution of Engineers (India) and the Young Scientist Award from VDGOOD Technology Factory, Kolkata. His research interests include air, water, and soil pollution; climate change; vulnerability; human health risk assessments; and GIS applications in environmental pollution and management. He is an active member of several reputed international professional bodies, including the IEI (Kolkata), ECI (New Delhi), AEACI (Mumbai), MSI (New Delhi), and IASTA (Mumbai).

Ramesh Singh earned his B.Tech in Civil Engineering with a Gold Medal from IT-BHU, Varanasi and M.S. in Sanitary Engineering from IHE Netherland. He worked as field engineer for 37+ years with Govt. of U.P. (India) and was involved with planning and implementation of various national and foreign aided projects for abatement works of rivers and lakes, and urban and rural water supply and environmental sanitation. He has worked as Visiting Faculty in Civil Engineering at IIT(BHU), Varanasi. Presently, he is engaged as Technical Advisor of Kokusai Kogyo Co. Ltd, Tokyo, Japan, and Directorate Environment and Climate Change, Govt. of Manipur (India). He has published a few research papers and is also a reviewer for reputed journals. He is fellow member of Indian Water Works Association (India), life member of Institution of Engineers (India), Indian Association of Environmental Management and The Indian Concrete Journal.

Prasant Kumar is a founding Co-Director of the pan-university Institute for Sustainability, Professor & Chair in Air Quality and Health, and founding Director of the internationally-leading research centre Global Centre for Clean Air Research (GCARE) at the University of Surrey, UK. He is the founder of successfully running Guildford Living Lab, a Trustee at Zero Carbon Guildford (ZERO), an Adjunct Professor at Trinity College Dublin, Ireland; and a Guest Professor at Southeast University, China. He joined the Senior Leadership team of the Faculty of Engineering and Physical Sciences as an Associate Dean (International) in 2020 and subsequently took up another senior leadership role to help set up the new pan-University Institute for Sustainability as a founding Co-Director in 2023. As an Associate Dean (International), he worked on implementing the University's Global Strategy and provided leadership to the international agenda, including student recruitment, transforming the PGR international placements through Turing and other mobility schemes, building several successful international research partnerships, coordinated numerous UG and PGT partnerships and joint PhD programmes, and led the QS submission for the University as a 'QS Impact Champion'. An engineer by training, Prof Kumar obtained his PhD (Engineering) from the University of Cambridge (UK) after winning a Cambridge-Nehru Scholarship and Overseas Research Scholarship award. He earned his master's degree in environmental engineering & management from the Indian Institute of Technology, Delhi, where he won the 'Outstanding Postgraduate Student Award' for his exemplary performance (CGPA 9.8/10 and rank 1). Prior to his PhD, he worked in the construction industry and at a CSIR Research Institute for about 7 years.

Contributors

Fanny Esther Berigüete Alcantara
Universitat Politècnica de Catalunya, Spain

Shaker Al-Hasnaawei
Department of Medical Analysis, Medical Laboratory Technique College, the Islamic University of Al Diwaniyah, Al Diwaniyah, Iraq

Geetanjali Amarawat
Swaminarayan University, Kalol, India

M.P. Akhtar
University of Bisha, Bisha, Saudi Arabia

Yacine Berrah
Echahid Cheikh Larbi Tebessi University, Tebessa, Algeria

Paolo Bellavista
University of Bologna, Italy

Dimpal Chauhan
Govt. College Jukhala, Bilaspur, India

Madderla Chiranjeevi
National Institute of Technology Karnataka, Surathkal, India
RV University, Bengaluru, India.

Christina Emmanouil
School of Spatial Planning and Development, Faculty of Engineering, Aristotle University of Thessaloniki, Thessaloniki, Greece

Nidhi Gupta
Baddi University of Emerging Sciences and Technology, Baddi, India

Debashisha Jena
National Institute of Technology Karnataka, Surathkal, India

Awanit Kumar
Career Point University, Kota, India

Kuldeep Kumar
Career Point University, Hamirpur, India

Prasant Kumar
University of Surrey, Guildford GU2 7XH, United Kingdom

Ravi Kumar
Baddi University of Emerging Sciences and Technology, Baddi, India

Shweta Kodihal
Methodist College of Engineering and Technology, Hyderabad, India

Md Mofazzal Hossain
Department of Law, Jagannath University, Dhaka, Bangladesh

Rohit Maheshwari
Career Point University, Kota, India

Satya Prakash Maurya
Madan Mohan Malaviya University of Technology, Gorakhpur, India

Qipei Mei
Civil and Environmental Engineering Department, University of Alberta, Edmonton, Alberta, Canada

Rahul Mishra
National Institute of Advanced Manufacturing Technology, Ranchi, India.

Sanjay Misra
Institute for Energy Technology, Norway

Tukaram Moger
National Institute of Technology Karnataka, Surathkal, India

Victor Babu Neela
Andhra University, Visakhapatnam, India

Pedro Ponce
Institute of Advanced Materials for Sustainable Manufacturing, Tecnologico de Monterrey, Monterrey, NL, Mexico

Peddada Jagadeeswara Rao
Andhra University, Visakhapatnam, India

N.G. Kalyani Reddy
G Pulla Reddy Engineering College, (Autonomous), Kurnool, 518007, India

Gazala Riyasat
Aligarh Muslim University, Aligarh, India

Richa Sharma
S.D.W.G. Govt. College Beetan, Una, India

Nirmal Singh
Sangam University, Bhilwara, India

Ramesh Singh
Indian Institute of Technology (Banaras Hindu University), Varanasi, India

Vijay P. Singh
Department of Biological and Agricultural Engineering &
Zachry Department of Civil & Environmental Engineering
Texas A&M University, College Station, Texas, United State

Swechchha Soni
Department of Biochemistry, Bundelkhand University, Jhansi, India

Zelina Sultana
Faculty of Business and Law, Anglia Ruskin University, Cambridge, United Kingdom
Department of Law, Jagannath University, Dhaka, Bangladesh

Sumiran Srivastava
Department of Biochemistry, Bundelkhand University, Jhansi, India

Ajay Kumar Suwalka
Sangam University, Bhilwara, India

Naveen Thakur
Career Point University, Hamirpur, India

Abhishek Verma
Madan Mohan Malaviya, Gorakhpur, India

Bhavana Verma
Delhi Technological University, New Delhi, India

Vijay Kumar Verma
Madan Mohan Malaviya University of Technology, Gorakhpur, India

Akhilesh Kumar Yadav
Chaoyang University of Technology, Taichung, Taiwan

Foreword

The contemporary era is marked by unprecedented environmental challenges that demand innovative, adaptive, and interdisciplinary solutions. Climate change, resource depletion, and recurring ecological crises are no longer confined to scientific debates; they have become pressing global concerns with direct implications for human well-being, economic stability, and planetary health. Addressing these challenges requires not only sound scientific understanding but also the integration of advanced modeling techniques with policy, governance, and societal action. This book, *Approaches of Adaptive Modeling for Sustainable Environmental Systems: Policy & Governance* arrives at a timely juncture. It brings together a diverse set of contributions that collectively emphasize the importance of adaptive frameworks in environmental system modeling. By integrating mathematical foundations, artificial intelligence, machine learning, and remote sensing with governance strategies, the volume underscores how science and technology can guide actionable policies in line with global sustainability goals.

What distinguishes this volume is its dual focus: it not only presents conceptual and methodological insights into adaptive modeling but also extends these approaches into practical applications for climate resilience, disaster management, and sustainable resource planning. The organization of the book into two sections—first, laying the groundwork of modeling approaches, and second, applying them within the context of global climate action—ensures that readers can appreciate both the theoretical underpinnings and the policy relevance of the subject matter. Equally noteworthy is the book's multidisciplinary outlook. By engaging experts from diverse fields, the editors have curated a volume that reflects the complexity of sustainability science. This holistic perspective is essential, for no single discipline can adequately address the interwoven challenges posed by climate change and environmental degradation.

I believe this volume will serve as a valuable reference for researchers, academicians, and practitioners, as well as a guide for policymakers seeking to integrate adaptive modeling into effective governance frameworks. It is also likely to inspire young scholars to engage in interdisciplinary research, contributing to the development of sustainable pathways for future generations. I commend the editors and contributors for their collective effort in producing this timely and important work. Their insights will not only enrich academic discourse but also provide practical guidance for advancing sustainability at local, national, and global levels.

Professor Vijay P. Singh, Ph.D., D.Sc., D. Eng. (HC, Waterloo), D.Sc. (HC, Guelph), Ph.D.

(HC, Basalicata), D.Sc. (HC, McGill), P.E., P.H., BC WRE (Hon.), D. EWRA (Hon.),
Dist.M. ASCE, Hon.Dist. M. IWRA, Dist.F. AGGS, Hon.M. AWRA, F. IAHR, F.AGU, EASA,
Academician (GFA), NAE
Distinguished Professor, Regents Professor
Caroline and William N. Lehrer Distinguished Chair in Water Engineering
Department of Biological and Agricultural Engineering &
Zachry Department of Civil & Environmental Engineering
Texas A&M University, College Station, Texas 77843-2117, United States

Preface

The increasing complexity of environmental challenges in the 21st century necessitates robust, adaptive, and scientifically informed approaches to ensure sustainability and resilience. Climate change, unsustainable patterns of resource utilization and recurring environmental hazards continue to impose significant risks on ecosystems, societies, and economies. In response, adaptive modeling has emerged as an indispensable framework, enabling the integration of scientific rigor, computational innovation, and policy mechanisms to address these multifaceted challenges. The present monograph, *Approaches of Adaptive Modeling for Sustainable Environmental Systems: Policy & Governance*, is an outcome of collaborative scholarship and aims to consolidate state-of-the-art research on adaptive environmental system modeling. It seeks to demonstrate how diverse methodologies including mathematical modeling, artificial intelligence, remote sensing, and machine learning can inform evidence-based decision-making, guide climate action strategies, and strengthen governance frameworks for sustainable development.

The book is organized into two complementary sections. The first section, *Essentials of Sustainable Environmental System Modeling*, provides conceptual and methodological foundations, covering mathematical approaches, AI-enabled monitoring and prediction, and the application of remote sensing technologies. It further elaborates on the principles and practices of adaptive modeling for climate change action plans. The second section, *Climate Change: Towards Sustainable Resource Planning and Management*, emphasizes applied perspectives by aligning modeling frameworks with global climate action agendas, particularly Sustainable Development Goal 13 (SDG-13). This section addresses disaster resilience, automation in policy and planning, AI-assisted decision-making during extreme events, and prospective frameworks for climate action up to 2030 and beyond. The volume brings together contributions from distinguished scholars and practitioners across disciplines, thereby reflecting the inherently multidisciplinary nature of sustainability science. It is envisioned that the book will serve as a valuable resource for researchers, academicians, policymakers, and professionals engaged in environmental governance, climate change mitigation, and sustainable resource management.

By advancing both theoretical insights and applied frameworks, this book aspires to bridge gaps between scientific innovation and policy implementation, ultimately contributing to the global endeavor of building adaptive and sustainable environmental systems.

Satya Prakash Maurya, Ph.D.
Akhilesh Kumar Yadav, Ph.D.
Ramesh Singh, M.S.
Prashant Kumar, Ph.D.

Acknowledgements

The editors wish to extend their sincere gratitude to all the contributors who have enriched this volume with their scholarly insights, empirical analyses, and innovative perspectives. Their commitment to advancing knowledge in the field of adaptive modeling and sustainable environmental systems has been instrumental in shaping the intellectual scope of this book. We also acknowledge with appreciation the reviewers and subject experts whose constructive feedback has helped refine the chapters and enhance the overall quality of the volume. Special thanks are due to the institutions and organizations that provided an enabling environment for research and collaboration. The support of universities, research centers, and professional networks has been invaluable in facilitating the exchange of ideas across disciplines and geographies.

The editors also recognize the contributions of the publisher for their guidance throughout the editorial and production process. Their cooperation has ensured the timely realization of this work. Finally, we express our gratitude to our families and colleagues, whose encouragement and support sustained us during the preparation of this volume. It is our sincere hope that the collective efforts embodied in this book will serve as a meaningful contribution to the literature on adaptive modeling, policy, and governance for sustainable environmental systems.

Satya Prakash Maurya, Ph.D.
Akhilesh Kumar Yadav, Ph.D.
Ramesh Singh, M.S.
Prashant Kumar, Ph.D.

Abbreviations

AI	artificial intelligence
ARIMA	autoregressive integrated moving average
BiLSTM	bidirectional long short-term memory
CNNs	convolutional neural networks
DL	deep learning
EIAs	environmental impact assessments
EEMD	ensemble empirical mode decomposition
EMD	empirical mode decomposition
FL	federated learning
GHI	global horizontal irradiance
GIS	geographic information system
GRUs	gated recurrent units
HI-SEAS	Hawai'i Space Exploration Analog and Simulation Research Station
IF	isolation forest
IMFs	intrinsic mode functions
KNNs	*K*-nearest neighbors
LSTM	long short-term memory
ML	machine learning
NITK	National Institute of Technology Karnataka
PE	permutation entropy
RFs	random forests
RS	remote sensing
SDGs	Sustainable Development Goals
SD	system dynamics

Part I

Essentials of Sustainable Environmental System Modeling

1

Current Directions of Sustainable Environmental Systems: Modeling Aspects

Gazala Riyasat, Zelina Sultana, and Md Mofazzal Hossain

1.1 Introduction

1.1.1 Sustainable Environmental Systems

Sustainability in environmental systems involves an integrated approach to human activity and natural resource management that maintains ecological integrity while serving present and future social needs. This introduction looks at the key concepts, difficulties, and frameworks of sustainable ecological management.

1.1.2 Core Concepts of Sustainability

Sustainability is driven by three pillars: environmental preservation, social equity, and economic viability. This "triple bottom line" approach recognizes that ecological health, human well-being, and economic advancement must all be balanced for systems to be truly sustainable. The preservation of natural capital or the collection of environmental assets that provide resources and ecosystem services essential to human survival and well-being is a special focus of environmental sustainability.

1.1.3 Environmental Systems Thinking

Environmental systems theory states that ecological and human systems are complex networks of interrelated pieces in which changes to one element affect others through feedback loops and emergent characteristics. From this viewpoint, we can better grasp how. The important functions of ecosystems include nutrient cycling, water purification, and climate management. Natural systems are affected by human activity. Interventions in one field may result in unexpected repercussions in other fields.

DOI: 10.1201/9781042013234-2

1.1.4 Current Environmental Challenges

Climate change, biodiversity loss, resource depletion, and pollution are interconnected environmental crises that require comprehensive management frameworks to address their cascading effects on global ecological stability. Integrated ecosystem management approaches recognize these challenges as systemically linked, emphasizing the management of entire ecosystems rather than isolated components through cross-boundary collaboration between stakeholders and governance levels. Circular economy models simultaneously address resource depletion and pollution by redesigning systems to eliminate waste, moving from linear "take-make-dispose" approaches to regenerative cycles that emphasize reuse, recycling, and product longevity while maintaining economic viability.

Nature-based solutions leverage natural systems to provide multiple co-benefits, such as wetland restoration, which simultaneously offers flood protection, carbon sequestration, biodiversity habitats, water filtration, and sustainable livelihoods, often providing more cost-effective alternatives than engineered alternatives. Adaptive management systems build flexibility in environmental governance by acknowledging uncertainty and emphasizing continuous monitoring, learning, and strategy adjustment based on outcomes and evolving scientific understanding. The planetary boundary framework provides science-based targets for maintaining Earth's stability by identifying nine critical Earth system processes and establishing safe operating spaces for humanity, thereby helping to prioritize interventions where dangerous tipping points are most imminent. The most effective sustainable environmental management requires the integration of these frameworks while simultaneously addressing social equity, economic viability, and political feasibility, alongside environmental goals. Success ultimately depends on coordinated action across multiple scales, from local community initiatives to international cooperation, recognizing that environmental challenges transcend boundaries and require systemic solutions that acknowledge the complex interdependencies between human societies and natural systems.

1.1.5 Several Frameworks Guide Sustainable Environmental Management

1.1.5.1 Circular Economy

The traditional "take-make-dispose" linear economic model is replaced with closed-loop systems that minimize waste and maximize resource efficiency through recycling, reuse, and regenerative design.

1.1.5.2 Ecosystem Services Valuation

To properly integrate natural system services into policy development and decision-making, it is important to acknowledge and measure their economic value (Daily et al., 2009).

1.1.5.3 Adaptive Management

Policies are viewed as experiments in this iterative approach, which emphasizes learning, monitoring, and modifying tactics in response to results and fresh data (Holling, 1978).

1.1.5.4 Planetary Boundaries

Nine essential Earth system processes are identified in this paradigm, each with thresholds that, if exceeded, might result in unacceptably altered environmental conditions. These include chemical contamination, biodiversity loss, and climate change (Steffen et al., 2015).

1.1.6 Implementing Sustainable Environmental Systems (Practical Implementation Typically Involves)

- Policy and regulatory frameworks that incentivize sustainable practices
- Technological innovation that increases efficiency and reduces environmental impacts
- Community engagement and participatory decision-making
- Education and knowledge sharing to build capacity and change behaviors
- Monitoring and assessment systems that track progress and inform adaptation

1.2 Role of Modeling in Sustainability

Modeling has become an essential tool for addressing sustainability issues in a variety of fields. Models offer crucial frameworks for comprehending complex systems, forecasting future events, and assessing possible responses as the world community struggles with climate change, resource depletion, and environmental degradation. This study examines the many facets of modeling involvement in sustainability science and practice, emphasizing important methods, uses, and difficulties.

1.3 Conceptual Frameworks for Sustainability Modeling

1.3.1 Systems Thinking and Complexity

Sustainability challenges are inherently complex and involve numerous interconnected social, economic, and environmental factors (Liu et al., 2015; Shirin et al., 2021). Systems thinking provides a foundational framework for modeling these complex relationships.

1.3.1.1 Complex Adaptive Systems

Sustainability challenges often exhibit properties of CAS, including non-linearity, emergence, and self-organization (Levin et al., 2013).

1.3.1.2 Socio Ecological Systems

The integration of social and ecological dimensions through frameworks such as the Social-Ecological Systems Framework developed by Ostrom (2009) enables more comprehensive modeling approaches.

1.3.2 Interdisciplinary Modeling Approaches

Sustainability modeling is a multidisciplinary approach that synthesizes diverse methodological frameworks and theoretical perspectives to address complex environmental and socioeconomic challenges. Integrated assessment models serve as comprehensive analytical tools that merge economic theory, earth systems science, and other disciplinary approaches to evaluate policy alternatives and forecast their potential consequences across multiple sectors and timeframes (Weyant, 2017). These models provide policymakers with quantitative assessments of the trade-offs between different sustainability pathways and their associated costs and benefits.

Agent-based models offer a complementary approach by simulating the behaviors and interactions of individual autonomous agents within complex systems, enabling researchers to observe how local-level decisions and interactions give rise to emergent system-wide phenomena and collective behaviors that may be difficult to predict using traditional top-down modeling approaches (Filatova et al., 2013). This bottom-up methodology is particularly valuable for understanding how individual choices regarding resource use, consumption patterns, and environmental behaviors aggregate to produce landscape-level sustainability outcomes.

Earth system models provide the biophysical foundation for sustainability analysis by representing the intricate physical, chemical, and biological processes that operate across the Earth's interconnected spheres, including atmospheric dynamics, oceanic circulation, terrestrial ecosystems, and cryosphere changes (Bonan & Doney, 2018). These models enable researchers to understand how human activities interact with natural systems and how these interactions influence long-term environmental stability and resource availability. Integrating these diverse modeling approaches creates a comprehensive analytical framework that captures both the human dimensions and biophysical constraints of sustainability challenges, providing essential tools for evidence-based decision-making in environmental management and policy development.

1.4 Applications for Modeling in Sustainability Domains

1.4.1 Climate Change Modeling

Climate modeling has emerged as a critical scientific tool for comprehending the complexities of climate change and developing robust projections of future environmental conditions and their societal implications. General circulation models represent the foundational framework for global climate analysis, using sophisticated mathematical representations of atmospheric and oceanic dynamics to simulate large-scale climate patterns and generate comprehensive projections of future climate scenarios under various greenhouse gas emission pathways and policy interventions. These models provide a scientific basis for understanding how global climate systems respond to anthropogenic forcing and natural variability across different temporal and spatial scales.

Downscaled regional climate models address the critical need for higher-resolution climate information by translating global model outputs into finer-scale projections that capture local topographic, land-use, and microclimatic influences on regional climate patterns (Giorgi & Gutowski, 2015; Singh et al., 2022). This enhanced spatial resolution is essential for developing targeted adaptation strategies and vulnerability assessments at the community and ecosystem levels, enabling stakeholders to understand how global climate change will manifest in specific geographic contexts and inform localized planning decisions.

Integrated assessment models specifically designed for climate policy evaluation, including prominent frameworks such as Dynamic Integrated Climate-Economy, Policy Analysis of the Greenhouse Effect, and Climate

Framework for Uncertainty, Negotiation, and Distribution, bridge the gap between climate science and economic analysis by quantifying the monetary costs of climate change impacts while simultaneously evaluating the economic effectiveness of various mitigation and adaptation strategies (Nordhaus, 2017). These policy-oriented models integrate climate projections with economic modeling approaches to provide decision-makers with comprehensive cost-benefit analyses of different policy options, helping to optimize resource allocation for climate action while balancing economic development goals with environmental protection objectives across multiple sectors and timeframes.

1.4.2 Resource Management Models

Sustainable resource management has become increasingly dependent on sophisticated modeling approaches that provide quantitative foundations for evidence-based decision-making across diverse natural resource sectors. Hydrological modeling systems serve as essential tools for water resource planning by simulating complex watershed processes, including precipitation patterns, groundwater dynamics, surface water flow, and water quality parameters, under alternative management scenarios and climate conditions (Loucks & van Beek, 2017). These models enable water managers to evaluate trade-offs between competing water uses, assess the sustainability of extraction rates, and develop adaptive strategies for maintaining water security in the face of changing precipitation patterns and increasing demand.

Forest management modeling frameworks provide comprehensive analytical capabilities for understanding forest ecosystem dynamics, including tree growth patterns, carbon storage potential, and the provision of diverse ecosystem services under different silvicultural practices and environmental conditions (Shifley et al., 2017). These models support forest managers in optimizing timber production while simultaneously maximizing carbon sequestration, biodiversity conservation, and other ecosystem services, thus enabling the development of management strategies that balance economic objectives with environmental sustainability goals across multiple forest types and management contexts.

Agricultural modeling systems have become indispensable tools for food security planning, using crop simulation models to predict agricultural productivity under varying climatic conditions, soil characteristics, and management practices (Rosenzweig et al., 2014). These models enable farmers, policymakers, and agricultural researchers to assess the vulnerability of food production systems to climate change, evaluate the effectiveness of different adaptation strategies, and optimize resource-use efficiency while maintaining agricultural sustainability. The integration of these diverse resource management modeling approaches creates a comprehensive analytical framework that supports sustainable resource use decisions across interconnected land, water, and agricultural systems.

1.4.3 Urban Sustainability Models

Urban climate models forecast urban heat island effects and other microclimate phenomena, land use change models simulate urban growth patterns and their environmental impacts, and urban metabolism models track the flows of energy, water, materials, and waste through urban systems (Kennedy et al., 2011). These models are examples of how they help address the unique sustainability challenges that urban areas present.

1.4.4 Energy System Models

Models such as energy system optimization models, which determine the least expensive routes for energy system transformation (DeCarolis et al., 2017); power system models, which model the production, transmission, and distribution of electricity with increasing proportions of renewable energy (Brown et al., 2018); and others are crucial to the shift to sustainable energy systems and energy-economy models, which capture interactions between energy systems and broader economic activities (Kriegler et al., 2015).

1.5 Advances in Modeling Approaches

1.5.1 Data-Driven Modeling

Sustainability modeling has evolved because of increasing quantities of data. For example, big data analytics integrate large, diverse datasets to provide a more thorough understanding of sustainability challenges (Hampton et al., 2013); remote sensing and Earth observation, which uses satellite and drone imagery to provide vital data for monitoring environmental changes and model validation (Turner et al., 2015); and machine learning and AI, which find patterns and relationships in complex data that traditional modeling approaches may ignore (Reichstein et al., 2019).

1.5.2 Participatory and Trans Disciplinary Modeling

Understanding the significance of diverse knowledge systems and stakeholder engagement, such as the coproduction of knowledge, increases the likelihood of model adoption by fostering trust, indigenous knowledge integration, which incorporates traditional ecological knowledge, and participatory modeling, which involves stakeholders in model development to improve model relevance and application (Voinov & Bousquet, 2010).

1.6 Challenges and Limitations

Some challenges and limitations are presented below.

1.6.1 Uncertainty and Model Complexity

There are some complexity and uncertainty, such as the balance between complexity and unpredictability, which states that while more intricate models may more accurately depict reality, they also become less transparent and intelligible to non-experts. Uncertainty, similar to describing and expressing uncertainty, quantification is still difficult but necessary for well-informed decision-making (Wilby & Dessai, 2010), and some validation of sustainability models is particularly challenging because of issues such as long-time horizons and complicated feedback (Oreskes et al., 1994).

1.6.2 Social and Institutional Dimensions

Value judgments in models inevitably embed assumptions and values that should be made explicit (Saltelli et al., 2020). Power dynamics, such as who builds models and for what purpose shapes their design and application, and the implementation gap, such as translating model insights into policy and practice, face numerous institutional barriers (Cash et al., 2003).

1.7 Future Directions

1.7.1 Integration of Multiple Models and Scales

There are several models and scales, such as Model Intercomparison Projects. Systematic comparisons of different models improve our understanding of their strengths and limitations (Eyring et al., 2016). Multiscale modeling, such as linking models across spatial and temporal scales, offers more comprehensive insights and model coupling, such as connecting models across domains (e.g., climate with agricultural or economic models) and capturing important feedback (van Vuuren et al., 2012).

1.7.2 Enhanced Decision Support

There is some planning, decision-making, and management, such as scenario planning, which uses models to explore multiple plausible futures rather than single predictions (Moss et al., 2010); robust decision-making, which identifies strategies that perform well across a range of scenarios (Lempert et al., 2013); and adaptive management, which uses models within iterative learning processes to refine strategies over time (Williams, 2011).

1.8 Conclusion

Modeling is essential for addressing sustainability issues because it enables us to comprehend intricate systems, predict future circumstances, assess possible solutions, and facilitate uncertain decision-making. Modeling techniques will become even more useful instruments for negotiating the path toward sustainability as they develop—incorporating more data, involving a wider range of stakeholders, and spanning several disciplines. However, when creating and using models, one must be mindful of their limits, underlying presumptions, and the social settings in which they are used. Technical proficiency is necessary for the best use of modeling for sustainability, but so are openness, inclusion, and dedication to bridging knowledge with action.

Acknowledgment

I sincerely thank my mentor and seniors for their invaluable guidance during this chapter. Special thanks to my peers for their advice and encouragement. I also value the tools and support provided by my institution has given me.

References

Bonan, G. B., S. C. Doney, 2018. Climate, ecosystems, and planetary futures: The challenge to predict life in Earth system models. *Science* 359:6375–8328. https://doi.org/10.1126/science. aam8328

Brown, T., D. Schlachtberger, A. Kies, S. Schramm, M. Greiner, 2018. Synergies of sector coupling and transmission reinforcement in a cost-optimized, highly renewable European energy system. *Energy* 160:720–739. https://doi.org/10.1016/j.energy.2018.06.222

Cash, D. W., W. C. Clark, F. Alcock, N. M. Dickson, N. Eckley, D. H. Guston, J. Jäger, R. B. Mitchell, 2003. Knowledge systems for sustainable development. *Proceedings of the National Academy of Sciences*, 100(14), 8086–8091. https://doi.org/10.1073/pnas.1231332100

Daily, G. C., S. Polasky, J. Goldstein, P. M. Kareiva, H. A. Mooney, L. Pejchar, T. H. Ricketts, J. Salzman, R. Shallenberger, 2009. Ecosystem services in decision making: Time to deliver. *Frontiers in Ecology and the Environment* 7(1):21–28. https://doi.org/10.1890/080025

DeCarolis, J., H. Daly, P. Dodds, I. Keppo, F. Li, W. McDowall, ... M. Zeyringer, 2017. Formalizing best practice for energy system optimization modeling. *Applied Energy* 194:184–198. https://doi.org/10.1016/j.apenergy.2017.03.001

Eyring, V., S. Bony, G. A. Meehl, C. A. Senior, B. Stevens, R. J. Stouffer, K. E. Taylor, 2016. Overview of the coupled model intercomparison project phase 6 (CMIP6) experimental design and organization. *Geoscientific Model Development* 9(5):1937–1958. https://doi.org/10.5194/gmd-9-1937-2016

Filatova, T., P. H. Verburg, D. C. Parker, C. A. Stannard, 2013. Spatial agent-based models for socioecological systems: Challenges and prospects. *Environmental Modeling & Software* 45:1–7. https://doi.org/10.1016/j.envsoft.2013.03.017

Giorgi, F., W. J. Gutowski, 2015. Regional dynamical downscaling and the CORDEX initiative. *Annual Review of Environment and Resources* 40:467–490. https://doi.org/10.1146/annurev-environ-102014-021217

Hampton, S. E., C. A. Strasser, J. J. Tewksbury, W. K. Gram, A. E. Budden, A. L. Batcheller, J. H. Porter, 2013. Big data and the future of ecology. *Frontiers in Ecology and the Environment* 11(3):156–162. https://doi.org/10.1890/120103

Holling, C. S., 1978. *Adaptive environmental assessment and management*. Wiley. https://pure.iiasa.ac.at/823

Kennedy, C., S. Pincetl, P. Bunje, 2011. The study of urban metabolism and its applications to urban planning and design. *Environmental Pollution* 159(8–9):1965–1973. https://doi.org/10.1016/j.envpol.2010.10.022

Kriegler, E., N. Petermann, V. Krey, V. J. Schwanitz, G. Luderer, S. Ashina, ... D. P. van Vuuren, 2015. Diagnostic indicators for integrated assessment models of climate policy. *Technological Forecasting and Social Change* 90:45–61. https://doi.org/10.1016/j.techfore.2013.09.020

Lempert, R. J., D. G. Groves, S. W. Popper, S. C. Bankes, 2013. A general, analytic method for generating robust strategies and narrative scenarios. *Management Science* 52(4):514–528. https://doi.org/10.1287/mnsc.1050.0472

Levin, S., T. Xepapadeas, A. S. Crépin, J. Norberg, A. De Zeeuw, C. Folke, ... B. Walker, 2013. Social-ecological systems as complex adaptive systems: Modeling and policy implications. *Environment and Development Economics* 18(2):111–132. https://doi.org/10.1017/S1355770X12000460

Liu, J., H. Mooney, V. Hull, S. J. Davis, J. Gaskell, T. Hertel, ... S. Li, 2015. Systems integration for global sustainability. *Science* 347(6225):1258832. https://doi.org/10.1126/science.1258832

Loucks, D. P., E. van Beek, 2017. *Water resource systems planning and management: An introduction to methods, models, and applications*. Springer. https://doi/10.007/978-3-319-44234-1

Moss, R. H., J. A. Edmonds, K. A. Hibbard, M. R. Manning, S. K. Rose, D. P. van Vuuren, ... T. J. Wilbanks, 2010. The next generation of scenarios for climate change research and assessment. *Nature* 463(7282):747–756. https://doi.org/10.1038/nature08823

Nordhaus, W. D., 2017. Revisiting the social cost of carbon. *Proceedings of the National Academy of Sciences* 114(7):1518–1523. https://doi.org/10.1073/pnas.1609244114

Oreskes, N., K. Shrader-Frechette, K. Belitz, 1994. Verification, validation, and confirmation of numerical models in the earth sciences. *Science* 263(5147):641–646. Https://doi.org/10.1126/science.263.5147.641

Ostrom, E., 2009. A general framework for analyzing sustainability of social-ecological systems. *Science* 325(5939):419–422. http://doi.org/10.1126/science.1172133

Reichstein, M., G. Camps-Valls, B. Stevens, M. Jung, J. Denzler, N. Carvalhais, Prabhat, 2019. Deep learning and process understanding for data-driven Earth system science. *Nature* 566(7743):195–204. https://doi.org/10.1038/s41586-019-0912-1

Rosenzweig, C., J. Elliott, D. Deryng, A. C. Ruane, C. Müller, A. Arneth, ... J. W. Jones, 2014. Assessing agricultural risks of climate change in the 21st century in a global gridded crop model intercomparison. *Proceedings of the National Academy of Sciences* 111(9):3268–3273. https://doi.org/10.1073/pnas.1222463110

Saltelli, A., A. Jakeman, S. Razavi, Q. Wu, 2021. Sensitivity analysis: A discipline coming of age. *Environmental Modelling & Software* 146: 105226. https://doi.org/10.1016/j.envsoft.2021.105226

Shifley, S. R., H. S. He, H. Lischke, W. J. Wang, W. Jin, E. J. Gustafson, ... J. Yang, 2017. The past and future of modeling forest dynamics: From growth and yield curves to forest landscape models. *Landscape Ecology* 32(7):1307–1325. http://doi.org/10.1007/s10980-017-0540-9

Shirin, S., A. K. Yadav, 2021. Assessment and suitability for the dumping of flyash in an abandoned mine. *Environmental Challenges* 4:100111. https://doi.org/10.1016/j.envc.2021.100111

Singh, V. P., S. P. Maurya, R. Singh, A. K. Yadav, 2022. An interdisciplinary modeling approach for dynamic adaptive policy pathways. In S. P. Maurya, A. K. Yadav, & R. Singh (Eds.), *Modeling and simulation of environmental systems: A computation approach* (pp. 343–353). CRC Press. https://doi.org/10.1201/9781003203445-22

Steffen, W., K. Richardson, J. Rockström, S. E. Cornell, I. Fetzer , E. M. Bennett, ... C. Folke, 2015. Planetary boundaries: Guiding human development on a changing planet. *Science* 347(6223):1259855. http://doi.org/10.1126/science.1259855

Turner, W., C. Rondinini, N. Pettorelli, B. Mora, A. K. Leidner, Z. Szantoi, ... C. Woodcock, 2015. Free and open-access satellite data are key to biodiversity conservation. *Biological Conservation* 182:173–176. https://doi.org/10.1016/j.biocon.2014.11.048

van Vuuren, D. P., L. Batlle Bayer, C. Chuwah, L. Ganzeveld, W. Hazeleger, B. van den Hurk, B. J. Strengers, 2012. A comprehensive view on climate change: Coupling of earth system and integrated assessment models. *Environmental Research Letters* 7(2):024012. http://doi.org/10.1088/1748-9326/7/2/024012

Voinov, A., F. Bousquet, 2010. Modeling with stakeholders. *Environmental Modeling & Software* 25(11):1268–1281. https://doi.org/10.1016/j.envsoft.2010.03.007

Weyant, J., 2017. Some contributions of integrated assessment models of global climate change. *Review of Environmental Economics and Policy* 11(1):115–137. https://doi.org/10.1093/reep/rew018

Wilby, R. L., S. Dessai, 2010. Robust adaptation to climate change. *Weather*, 65(7): 180–185. https://doi.org/10.1002/wea.543

Williams, B. K., 2011. Adaptive management of natural resources—framework and issues. *Journal of Environmental Management* 92(5):1346–1353. https://doi.org/10.1016/j.jenvman.2010.10.041

2

Adaptive Modeling for Sustainable Environmental Systems: Integrating Traditional Approaches with Modern Innovations

Swechchha Soni, Sumiran Srivastava, Christina Emmanouil, and Akhilesh Kumar Yadav

2.1 Introduction

In an era marked by rapid environmental change, increasing climate volatility, and complex socio-ecological interactions, sustainable environmental planning has become increasingly critical and challenging. Traditional environmental modeling approaches such as system dynamics (SD), static simulations, and environmental impact assessments (EIAs) have played a foundational role in informing environmental policy and management decisions. However, these methods often struggle to respond to the unpredictable and dynamic nature of current ecological and societal systems. As environmental systems become increasingly interconnected, static or linear models risk producing outdated or inflexible insights. Moreover, advancements in digital technologies, real-time data access, and participatory governance offer opportunities to improve our understanding of and response to environmental change. This has led to adaptive modeling, a methodology that incorporates learning, feedback, and iterative decision-making into modeling processes.

Adaptive modeling is not intended to replace traditional approaches but rather enhances them by integrating conventional knowledge with modern tools and collaborative practices. This integration allows for more resilient, responsive, and context-specific environmental strategies that can guide sustainable development in both the short and long term. This study explores the conceptual evolution of adaptive modeling, reviews the literature on both traditional and modern methods, and highlights how their integration can strengthen sustainability planning. Through case studies and policy analysis,

 DOI: 10.1201/9781042013234-3

key strategies, tools, and institutional mechanisms that support the transition toward adaptive and inclusive environmental governance are identified.

2.1.1 Background

The growing complexity of environmental degradation, climate insecurity, and socio-ecological change has increased the demand for flexible and adaptive sustainable planning tools. Traditional modeling methods, such as SD, stationary land-use simulations, and EIAs, have played a central role in environmental decision-making. Nevertheless, these models frequently operate on fixed parameters and assume system stability, thereby limiting their effectiveness in dynamic environments. (Van Eynde et al., 2024; Wang & Ren, 2025). To overcome these limitations, adaptive modeling has emerged as a promising solution. Adaptive modeling integrates feedback mechanisms, real-time data, and stakeholder participation, allowing models to evolve as conditions change. This dynamic capability enhances the applicability and delicacy of policy and planning opinions under the conditions of inquiry (Luleci et al., 2024; Santos & Carvalho, 2025; Singh et al., 2022).

Technological advances, such as machine learning (ML), Geospatial Artificial Intelligence (GeoAI), and cloud-based geographic information system (GIS) systems, now enable adaptive models to incorporate massive datasets, process them in real time, and simulate outcomes under different scenarios. These tools make it possible to model socio-ecological systems more realistically by capturing nonlinear dynamics, sudden shifts, and spatial variability (Mai et al., 2025; Koldasbayeva et al., 2024; Shirin at al., 2021). Recent studies have also supported mongrel modeling approaches that combine traditional system-dynamic fabrics with ultramodern artificial intelligence (AI)-driven and participatory styles. These intertwined models aim to maintain theoretical robustness while enhancing rigidity and responsiveness to real-world complications. (Zhang et al., 2024; Supradianto Nugroho & Uehara, 2023). Understanding and advancing these integrated approaches are critical for future sustainability efforts. Adaptive modeling offers not only a methodological shift but also a systemic shift bridging technological innovation, stakeholder engagement, and environmental resilience.

2.1.2 Adaptive Modeling Approaches in Sustainable Environmental Systems

Traditional environmental models, such as static simulations or classic SD models, often assume fixed relationships and one-time scenarios. Such models can be too "static, centralized, and closed" to easily incorporate new data or reuse components. For example, Wang and Ren (2025) noted that a static scenario approach fails to address the rapidly changing nature of urban environments, arguing that "real-time or adaptive modeling would make the framework more relevant for continuous, long-term planning." Similarly, critics have observed that many mainstream socioeconomic–environmental models simplify complex trade-offs. F. Demaria et al. (2024) review dozens of integrated environment–society–economy models and find several key limitations: they reduce climate mitigation to cost–benefit formulas, ignore overshoots of planetary boundaries, rely on overly optimistic technology assumptions, and neglect sociotechnical transition challenges. In summary, traditional models tend to oversimplify feedback, omit stakeholder inputs, and assume stable, single-policy futures gaps that can mislead decision-makers under uncertainty and change.

- **Feedback and Dynamics:** Classic SD models capture feedback loops and have been useful for policy insights; however, by aggregating variables, they often oversimplify fine-scale complexity. For example, SD models can neglect local heterogeneity and suffer from limited time-series data. They may also inadequately represent unexpected shocks or shifting baselines. In short, purely static or linearized models lack the flexibility to adjust to new information.
- **One-off Scenarios:** Many conventional tools run off-line "once-ahead" simulations (e.g., single-run climate or hydrology forecasts) that cannot easily incorporate real-time data streams or participatory inputs. As Luleci et al. (2024) argued, moving beyond these static models is essential: static connectivity or land-use models have only "limited usefulness for decision-making" when future land changes are uncertain.
- **Critiques by scholars:** Several researchers have highlighted theselimitationss. For example, Van Eynde et al. (2024) emphasize the failure of integrated assessment models. Others have similarly faulty conventional climate–economy models for neglecting equity, nonmarket values, and systemic feedback. Such critiques motivate the search for more adaptive or dynamic approaches (Figure 2.1).

2.1.3 Rise of Adaptive Modeling

In response to these limitations, a growing body of research has advocated adaptive modeling approaches. Adaptive modeling is characterized by feedback loops, real-time data integration, and iterative stakeholder involvement. Models are continuously updated as new observations arrive, and they explicitly account for the interactions between natural and human systems.
The key principles include the following:

- **Feedback and self-adjustment:** Adaptive models incorporate continuous feedback from the environment. For example, some urban planning tools have been designed to update scenario evaluations as new sensor data or policy outcomes have become available. Rather than

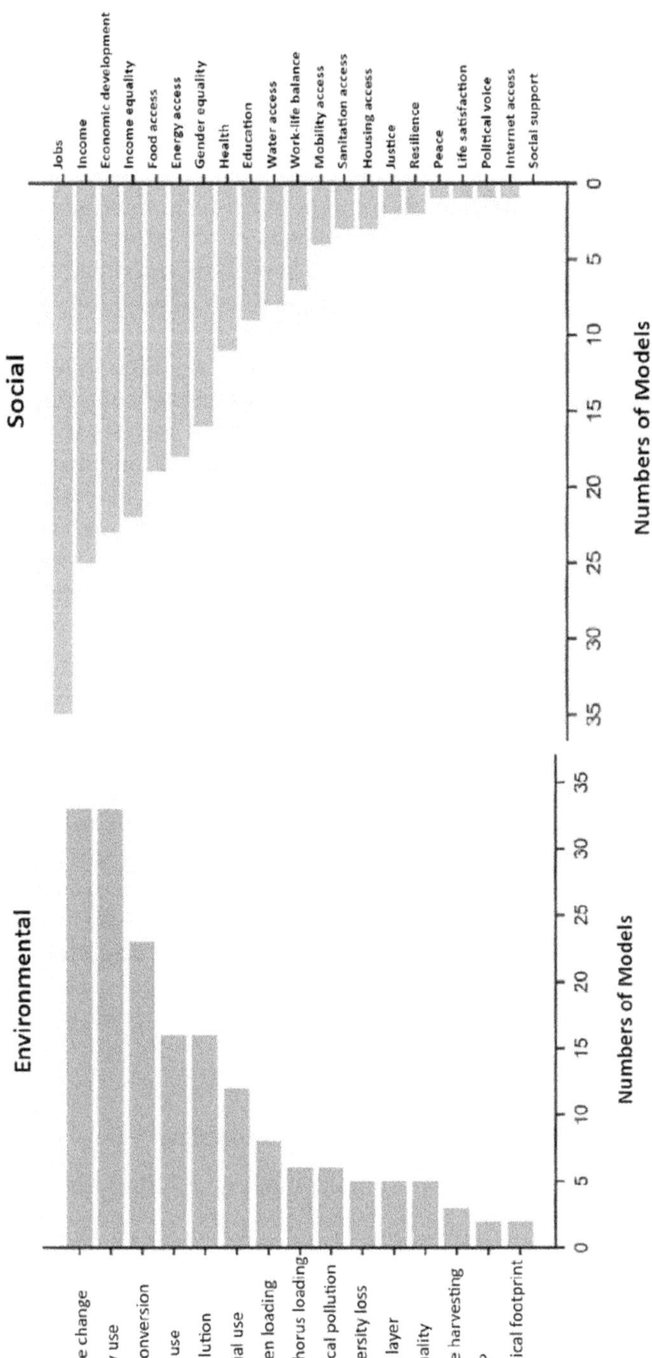

FIGURE 2.1
Coverage of environmental and social indicators across the medium list of 50 models.

static scenario runs, these systems loop model outputs back into inputs, adjusting parameters, or scenarios in response to observed changes.

- **Real-time data streams:** Innovations in sensing (IoT, remote sensing [RS]) and cloud platforms enable near-real-time data feeds. GeoAI techniques now fuse satellite images, ground sensors, and citizen science data into models. GeoAI-driven Earth-system models, for example, leverage vast geo-referenced datasets to refine short-term forecasts and long-term climate projections (D. Koldasbayeva et al., 2024). In practice, this means updating hydrological models with recent streamflow measurements, or revising biodiversity models when new survey data are available.
- **Participatory and iterative design:** Adaptive modeling often involves stakeholders directly. Participatory modeling engages community members, managers, and scientists in the code sign of models so that local knowledge and values shape the model structure and use. In such approaches, models are coproduced iteratively: stakeholders help frame the problem, prioritize variables, and test scenarios, while modelers update the model on the basis of that feedback (analogous to Figure 2.2, techniques in Hubacek et al., 2017, although cited sources are mostly pre-2021). Recent reviews (e.g., Santos & Carvalho, 2025) highlight how AI-driven tools can foster adaptive management strategies by integrating stakeholder input and domain knowledge, enhancing collaboration toward sustainability goals.
- **Adaptive management cycle:** The adaptive model is embedded in a management cycle: Plan – Implement – Monitor – Adjust. This aligns with the classic "adaptive management" philosophy but is now powered by computing. Models help test interventions in silico, managers monitor outcomes, and then revise the model. For example, Pokhrel's "human-water" hydrologic models (described in MSU news 2024) adapt water cycle simulations based on changing irrigation or extraction patterns. Although this press piece is not a peer-reviewed source, it exemplifies the idea of adaptive hydrological modeling that responds to human feedback (Table 2.1).

2.1.4 Technological Innovations Enabling Adaptive Modeling

Several recent technological advances have underpinned the shift to adaptive models:

- **AI and ML:** Modern AI/ML methods can detect complex, nonlinear patterns from large environmental datasets. *Data-driven geospatial models* offer "unique opportunities for environmental monitoring due to domain and scale adaptability and computational efficiency". In

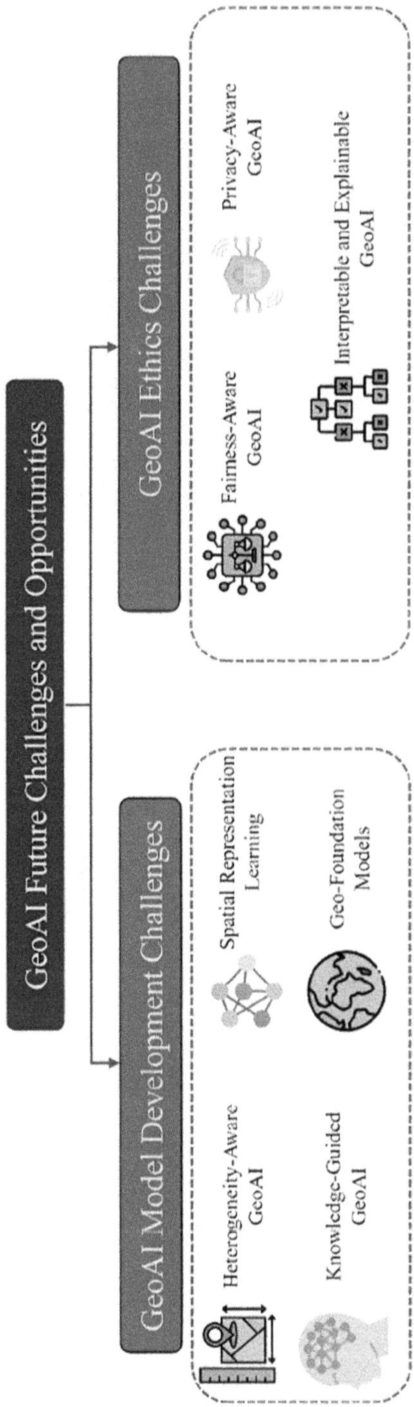

FIGURE 2.2
Future challenges and opportunities for geospatial artificial intelligence.

TABLE 2.1

Criteria on a continuum from traditional to alternative approaches to model-based decision support

Criteria	'Traditional' Approaches	'Alternative' Approaches
Degree of stakeholder participation	Individuals: stakeholders are individually consulted for their inputs	Groups: interactive team learning situation where the model is almost a byproduct of the modeling process
Timing of participation	At the beginning and/or the end of the process	Throughout the process
Role of stakeholders and scientists	Stakeholder is seen as data provider and as lay-person whereas the modeler is seen as expert	Stakeholders are seen as cocreator of knowledge; the modeler is seen as one of the stakeholders; multiple legitimate knowledge sources exist
Framing the research	Stakeholders work with given questions/problems and within a given frame	Stakeholders guide problem formulation and help design the conceptual frame of the model
Purpose	To develop better models	Multiple purposes: I) Improve model; II) group learning/capacity building – improves acceptance of outcome; leads to consensus – increases stakeholder buy-in and improves implementation; III) transparency; IV) legitimacy
Outcomes	Predictable outcomes	Less predictable outcomes as stakeholders shape the process
Power sharing	Little	Potentially high
Model integration	Models of natural and human world are separate	Coupled models of natural-human dynamic systems
Validity of models	Scientific standards as metrics of success of the model	Untraditional metrics of success of the process, e.g. measuring new knowledge, evaluating adaptive behavioral, social learning and institutional change

practice, ML algorithms (e.g., neural nets and random forests) have been integrated into environmental models to capture factors that are difficult to parameterize physically. For example, Zhang et al. (2024) fused SD models with land-change models and ML to simulate ecosystem carbon sequestration under climate scenarios. The results achieved high accuracy by allowing ML to handle the nonlinear influences of socioeconomic and climate factors on net primary productivity. Similarly, GeoAI applications now enhance Earth-system models. Mai et al. (2025) described how *GeoAI-enabled* climate models use satellite and sensor data to improve both short- and long-term climate projections. In short, AI/ML tools make it feasible to update models rapidly when new data arrive and extract insights from heterogeneous sources.

- **GIS and RS:** Advances in high-resolution spatial data have increased model adaptability. Continuous streams of satellite imagery and ground-sensor networks feed dynamic GIS layers. For example, real-time air quality or land cover maps can be used to drive urban models, whereas LiDAR and drones can be used to enrich habitat models. As Koldasbayeva et al. (2024) noted, geospatial modeling tools (ML and DL) overcome many limitations of pure equation-based models by efficiently handling diverse spatial data. GIS platforms have become more powerful and cloud-based, enabling stakeholders to visualize and tweak models on the fly (e.g., digital twins of cities or ecosystems).
- **Cloud computing and IoT:** The scalable computing power of cloud platforms makes it practical to run complex models continuously or at high frequency. Environmental "digital twins" – virtual replicas of cities, watersheds, or the entire built environment – often run on cloud servers, integrating streams from IoT devices. For example, Luleci et al. (2024) described a "Community Twin Ecosystem" built on a game engine and cloud technology; it captures dynamic, multidimensional community data and offers real-time scenario simulation via an easy interface. Similarly, national weather and climate centers use cloud supercomputing to assimilate incoming sensor data into forecasts several times per day. This technological convergence means that models can adapt as if online, ingesting new inputs and delivering updated outputs constantly.
- **Participatory platforms and decision support tools:** New software and online tools (often integrating ML and GIS) make participatory modeling more accessible. These platforms allow experts and non-experts to run simulations jointly, adjust parameters, and observe the impacts of interventions. Innovations in UX (e.g., interactive dashboards, apps) lower the barrier to "citizen science" contributions, effectively embedding stakeholder feedback into model workflows.

Together, these innovations mean that models are no longer siloed desktop applications but are part of interconnected cyber-physical systems. They can leverage AI, cloud computing, and geospatial data as self-updating decision-making aids.

2.1.5 Integrating Traditional and Modern Approaches

Rather than discarding older models entirely, many researchers now advocate hybrid approaches that combine the strengths of both worlds. For example, hybrid SD–ABM models merge SD (macro-level, aggregate behavior, and feedback) with agent-based models (micro-level, individual, or local interactions). Such hybrids can capture high-level trends while still representing heterogeneous agents, albeit at the expense of greater complexity. Supradianto Nugroho and Uehara (2023) review this trend, noting that as demand grows for models capturing both macro- and micro-SES (social–ecological system) dynamics, hybrid models are increasingly used, despite the challenge of greater conceptual overhead.

Other studies fused physics-based models with AI. In hydrology and climate science, scholars are layering ML emulators on top of traditional process models, using AI to calibrate or accelerate the physical model rather than replacing it entirely. For example, Mai et al., 2025 showed that deep learning could capture sub-grid atmospheric processes and inform global climate models, effectively optimizing traditional process-based ESMs for better projections.

Zhang et al. (2024) demonstrated a concrete hybrid: they used a SD framework for carbon cycle flows, a cellular automata land-change model for urbanization, and ML to link climate/emissions scenarios to ecosystem productivity. The result is a flexible model that respects system theory (via SD) but is tuned by data-driven ML where needed. Overall, there is a trend toward modular interoperable modeling frameworks. Initiatives such as the "Model as a Service" concept propose making models accessible as web services so that multiple models can be chained or swapped dynamically. In practice, this means that a water model could call a land-use model in real time or plug in an updated AI component when new data arrives.

2.2 Conclusion

In summary, traditional one-shot models give way to iterative data-assimilating approaches. Adaptive modeling combines continuous data inputs, feedback loops, stakeholder coproduction, and advanced computing. It leverages GIS, AI/ML, cloud computing, and digital twins to create models that learn and evolve. Integrating the old and new, for example, embedding AI modules in classic process models or linking SD and agent-based components is a promising strategy to retain theoretical grounding while gaining flexibility. The literature since 2021 reflects this shift, with authors reporting on hybrid frameworks, criticizing outdated methods, and demonstrating new AI–GIS–cloud tools for sustainability modeling.

Acknowledgement

The authors express their gratitude to the reviewers and editor for dedicating their valuable time to enhancing this work.

References

Demaria, F. (2024). Modeling what matters: How do current models handle environmental limits and social outcomes. *Journal of Cleaner Production*, 476, 143777. https://doi.org/10.1016/j.jclepro.2024.143777

Hubacek, Klaus, , Ravera, F., Tarrasón, D., Prell, C. (2017). Participatory modeling. *In International Encyclopedia of Geography: People, the Earth, Environment and Technology (eds D. Richardson, N. Castree, M.F. Goodchild, A. Kobayashi, W. Liu and R.A. Marston)*. https://doi.org/10.1002/9781118786352.wbieg0293

Koldasbayeva, D., Tregubova, P., Gasanov, M. et al. (2024). Challenges in data-driven geospatial modeling for environmental research and practice. *Nature Communications*, 15, 10700. https://doi.org/10.1038/s41467-024-55240-8

Luleci, F., Sevim, A., Ozguven, E. E., Catbas, F. N. (2024). Community twin ecosystem for disaster resilient communities. *Smart Cities*, 7(6), 3511–3546. https://doi.org/10.3390/smartcities7060137

Mai, G., Xie, Y., Jia, X., Lao, N., Rao, J., Zhu, Q., Liu, Z., Chiang, Y., Jiao, J. (2025). Toward the next generation of geospatial artificial intelligence. *International Journal of Applied Earth Observation and Geoinformation*, 136, 104368. https://doi.10.1016/j.jag.2025.104368.

Nugroho, S., Uehara, T. (2023). Systematic review of agent-based and system dynamics models for social-ecological system case studies. *Systems*, 11(11), 530. https://doi.org/10.3390/systems11110530

Santos, M. R., Carvalho, L. C. (2025). AI-driven participatory environmental management: Innovations, applications, and future prospects. *Journal of Environmental Management*, 373, 123864.

Shirin, S., Yadav, A. K. (2021). Assessment and suitability for the dumping of flyash in an abandoned mine. *Environmental Challenges*, 4, 100111. https://doi.org/10.1016/j.envc.2021.100111

Singh, V. P., Maurya, S. P., Singh, R., Yadav, A. K. (2022). An interdisciplinary modeling approach for dynamic adaptive policy pathways. In S. P. Maurya, A. K. Yadav, & R. Singh (Eds.), *Modeling and Simulation of Environmental Systems: A Computation Approach* (pp. 343–353). CRC Press. https://doi.org/10.1201/9781003203445-22

Van Eynde, R., Greenford, D. H., O'Neill, D. W., Demaria, F. (2024). Modelling what matters: How do current models handle environmental limits and social outcomes? *Journal of Cleaner Production*, 476, 143777. https://doi.org/10.1016/j.jclepro.2024.143777

Wang, Z., Ren, F. (2025). Developing a decision support system for sustainable urban planning using machine learning-based scenario modeling. *Scientific Reports*, 15, 13210. https://doi.org/10.1038/s41598-025-90057-5

Zhang, Y., Zhang, Y., Yang, J., Wu, W., Tao, R. (2024). Integrating system dynamics, land change models, and machine learning to simulate and predict ecosystem carbon sequestration under RCP-SSP scenarios: Fusing land and climate changes. *Land*, 13(11), 1967. https://doi.org/10.3390/land13111967

3

Intervention of Artificial Intelligence in Sustainable Environmental System Monitoring and Prediction

Abhishek Verma, Bhavana Verma, Yacine Berrah, and Pedro Ponce

3.1 Introduction

During the last few decades, the unusual extent of environmental degradation has emerged as one of the most significant challenges to human society, posing far-reaching consequences for ecosystem health, biodiversity preservation, and human well-being. The environmental crises of global warming, biodiversity loss, plastic pollution, and forest depletion have all reached critical thresholds that require innovative new technologies that incorporate corrective interventions into their complex interdependencies. Environmental systems are complex, interrelated, and difficult to monitor because of the considerable limitations in accuracy, breadth, and near-time usefulness of traditional monitoring programs. Furthermore, environmental monitoring programs relying on traditional systems require expensive and specialized equipment and experts to operate them, suffer from losses in operational functionality owing to human error, and cannot be used to make predictions regarding environmental trends and their immediate approaches toward unmanageability, as illustrated in **Figure 3.1**.

With the advent of artificial intelligence (AI) technologies, the availability of environmental evidence and various forms of environmental information, the use of satellite observations, concurrent access to sensor networks via the Internet of Things (IoT), and observable ground stations continue to expand exponentially for environmental scientists and decision-makers. Incompatible sources of data can help provide unprecedented opportunities to correlate massive amounts of interdependent, heterogeneous, environmental data collected from different sources, such as satellite observations, meteorological data, IoT sensor networks, and citizen

DOI: 10.1201/9781042013234-4

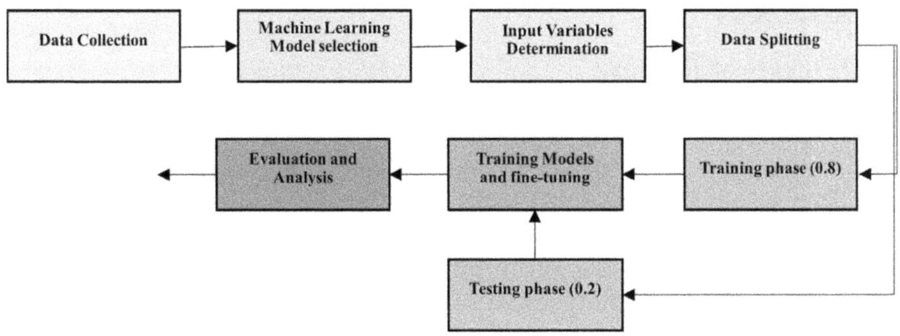

FIGURE 3.1
Basic steps for developing a prediction model.

science data campaigns made available to facilitate learning by users and provide valuable additions to scientific inquiry, if not, contemporaneous data values. These data sources cover a complete range of possible environmental indicators, such as air quality indices, water pollution levels, rates of forest depletion, and wildlife populations, and establish patterns of mitigation activity. New algorithms will be able to analyze these variables in real time and make correlations that would have been impossible from investigations by human means. New AI algorithms can examine these variables in real time, making available insights that would be impossible to discover manually while simultaneously offering predictive capacity for environmental change forecasting.

AI systems can anticipate and track environmental shifts and provide information to act against environmental hazards using machine learning (ML), neural networks, and deep learning (DL) algorithms. By collecting and interpreting environmental data, AI-powered systems can help maximize conservation initiatives and sustainable resource use to identify harmful agents in the environment and protect human health and ecosystems as technology advances. AI represents a potential component for designing sensors to recognize and understand these harmful agents. AI-based environmental monitoring systems can expand the decision support capacity of environmental managers, policy planners, and conservation practitioners working to manage specific environmental concerns. Environmental data calculated via AI can be important for sound decision-making if they are sufficiently analyzed and visualized. Modern AI techniques and tools have been proposed to analyze the growing volumes of environmental data in integrated and comprehensive forms, enabling real-time monitoring, predictive modeling, and automated alert systems (Wang et al., 2025; Rastogi et al., 2025).

3.2 Literature Review

3.2.1 Artificial Intelligence in Environmental Monitoring

Early applications of deep recurrent neural networks demonstrated enhanced spatiotemporal forecasting capabilities for air pollution, such as achieving accurate short-term $PM_{2.5}$ predictions (Fan, 2017) and improving the accuracy of $PM_{2.5}$ estimates in urban regions (Pak et al., 2020). Physics-informed frameworks combine chemical transport models with data-driven AI to embed fundamental physical laws into the neural architecture, enhancing model generalization, and ensuring physically plausible outputs (Kumar et al. 2022; Patel et al., 2022).

3.2.2 Hybrid Artificial Intelligence Models for Pollutant Prediction

Ensemble modeling approaches that integrate satellite, ground-level, and meteorological data have yielded robust NO_2 exposure estimates, significantly reducing health risk assessment uncertainties(Huang et al., 2022). The three-stage modeling framework developed by Huang *et al.* demonstrates how ensemble models can effectively predict daily NO_2 concentrations with high spatiotemporal resolution, achieving R^2 values of 0.89, with an RMSE of 5.62 μg/m³(*An Ensemble Model-Based Estimation of Nitrogen Dioxide in a Southeastern Coastal Region of China*, n.d.). Additionally, a multiresolution ensemble model utilizing generalized additive approaches to combine three decision tree algorithms achieved a tenfold cross-validated R^2 of 0.83 for 1 km resolution prediction. Similarly, transformer-based architectures have been adapted to capture long-range dependencies in NO_x time series, leading to marked improvements over standard LSTM models (Rai et al., 2023) and achieving superior performance, with the mean absolute error (MAE) decreasing from 0.29625 to 0.19829 when positional encoding is incorporated (Sonata & Heryadi, 2024); in contrast, the hybrid RF-S2S model, which combines random forest (RF) feature selection with sequence-to-sequence networks, achieves a Pearson correlation coefficient of 0.93 for NO_2 prediction, outperforming traditional LSTM approaches (Jia et al., 2023).

3.2.3 Data Fusion and Multimodal Integration

Multisource fusion techniques that leverage auxiliary variables, such as aerosol optical depth and land-use parameters, have effectively addressed the weak correlation between satellite-retrieved column NO_2 and surface concentrations, improving the spatial coverage of ground-level estimates. Hybrid DL models augmented with metaheuristic optimization algorithms have further reduced the forecasting error through adaptive hyperparameter tuning, demonstrating gains in both accuracy and stability. Virtual observation networks constructed via gradient boosting machines combined with dense meteorological station data have provided high-density $PM_{2.5}$ predictions nationwide, offering a scalable approach for regions with sparse air quality monitoring.

3.3 Case Studies in Artificial Intelligence-Driven Environmental Monitoring

The implementation of AI in environmental monitoring illustrates the practical application of hybrid AI architecture for environmental prediction, highlighting that both the BREATH-Net framework and CATALYST system represent significant advances in computational environmental science, demonstrating how sophisticated neural architectures can address complex pollution monitoring challenges in urban environments.

3.3.1 Tools and Technologies for Artificial Intelligence-Driven Environmental Monitoring

3.3.1.1 Artificial Intelligence and Machine Learning Frameworks

The terrain of AI architectures for environmental purposes offers environmental modeling and prediction activities. TensorFlow 3.0, created by Google Brain, is a significant new development in ML infrastructure with multi-GPUs, TPUs, and machine clusters, demanding extensive-scale computational power for end-to-end ML pipelines and TensorFlow Federated for privacy-preserving environmental data analysis at multiple institutions.

PyTorch has undergone widespread adoption and is now well suited for experimental environmental monitoring tasks with open neural network exchange (ONNX), allowing for easy transfer of models between various frameworks across multiple computing environments.

Scikit-learn continues to be the de facto library for common ML methods in environmental science, offering solid algorithm implementations, including RF support vector machines and ensemble algorithms. Its compatibility with other Python packages renders its workflow for environmental data analysis seamless.

3.3.1.2 Environmental Data Processing and Analysis Tools

NumPy and Pandas, the two main tools for data manipulation in Python, come into play when the data involves satellite imagery, climate time series, and sensor networks, which address large environmental datasets and involve computations with many variables. They supply functionalities for data cleaning and transformation, along with time-series analysis. Specialized SciPy functions complement NumPy for image processing required in the analysis of satellite images and remote sensing (RS) applications and assist in environmental prediction model parameter estimation.

Matplotlib and Seaborn provide great visualization opportunities, plotting, mapping, and interactive visualizations, which are much needed for communicating environmental discoveries to stakeholders and policymakers.

3.3.1.3 Satellite Data Processing and Remote Sensing Tools

Pysat (Python Satellite Data Analysis Toolkit) provides a comprehensive framework for satellite data analysis from beginning to end, including downloading, loading, cleaning, managing, processing, and analyzing data, and xarray integration ensures the consistent handling of diverse scientific measurements while supporting metadata standards consistent with netCDF CF-1.6.

Google Earth Engine (GEE) represents a revolutionary cloud-based platform for planetary-scale geospatial analysis [14] with access to petabytes of satellite imagery and geospatial datasets, enabling researchers to perform complex environmental analyses without requiring local computational infrastructure and supporting applications ranging from deforestation monitoring to climate change analysis.

QGIS and ArcGIS provide sophisticated geographic information system (GIS) capabilities for environmental analysis, help scientists and planners understand potential environmental issues, and offer open-source alternatives with extensive plugin ecosystems that support ecological modeling and analysis workflows.

3.3.1.4 Internet of Things and Sensor Network Platforms

IoT-based environmental monitoring systems have transformed real-time environmental data collection through distributed sensor networks for sensor interfacing and Raspberry Pi devices, thereby creating scalable monitoring networks to secure data transmission.

The **Arduino and Raspberry Pi ecosystems** provide accessible platforms for developing custom environmental monitoring solutions, creating comprehensive environmental monitoring solutions spanning from data collection to analysis and visualization.

Commercial IoT platforms such as AWS IoT, Microsoft Azure IoT, and Google Cloud IoT provide enterprise-grade infrastructure for large-scale environmental monitoring deployments, ranging from small research projects to national monitoring networks.

3.3.1.5 Cloud Computing and Big Data Platforms

Cloud-based GIS solutions have revolutionized environmental data accessibility and collaboration. Platforms such as Esri's ArcGIS Online, Microsoft Azure, and AWS Location Services enable organizations to store, share, and analyze spatial data without extensive on-premise infrastructure. Cloud computing facilitates real-time collaboration among distributed research teams while providing scalable computational resources for large-scale environmental analyses.

Big data frameworks, such as Apache Spark and Hadoop, support the processing of massive environmental datasets that exceed local computational capabilities. These frameworks enable parallel processing of satellite imagery, climate model outputs, and sensor network data. Integration with ML libraries creates powerful platforms for environmental big data analytics.

Environmental data APIs provide programmatic access to diverse datasets. including the Pollen API, Air Quality API, and Solar API, which help with decision-making and public awareness applications.

3.3.1.6 Specialized Environmental Monitoring Tools

iMESc (interactive ML app for Environmental Sciences) represents a new generation of user-friendly, integrated, supervised, and unsupervised ML methods for enhancing reproducibility to ensure seamless transitions between analytical workflows.

Spectroscopy and sensor integration tools combine advanced chemical analyses with flexible substrates to enable the detection of trace pollutants, explosives, and industrial contaminants under field conditions. Electronic nose (E-nose) technologies utilize ML algorithms to identify toxic compounds through chemical signature recognition.

Predictive maintenance and monitoring systems utilize AI to ensure the continuous operation of monitoring networks. Predictive analytics reduce downtime and maintenance costs while maintaining data quality in long-term environmental monitoring programs.

3.3.1.7 Database and Data Management Systems

Cloud database systems provide scalable infrastructure for environmental data storage and management for distributed data processing, as well as high efficiency, scalability, and fault tolerance, which are essential for managing massive datasets generated by modern environmental monitoring systems.

Time series databases specializing in environmental monitoring applications handle the unique characteristics of environmental data, including irregular sampling intervals, missing data, and multiparameter measurements. These systems provide optimized storage and retrieval for long-term environmental datasets, while supporting real-time analytics and alert capabilities.

3.3.2 Case Study 1: BREATH-Net Framework for NO_2 Prediction in Delhi

3.3.2.1 Study Context and Environmental Significance

Delhi is one of the most challenging urban air quality scenarios because it contains several sources of pollution, including industrial activities, traffic, and biomass consumption. Nitrogen dioxide is one of the concerns considered in the context of meteorological variability and patterns. Figure 3.2 illustrates the BREATH-Net framework. The combination of Bi-LSTM and transformer-based attention mechanisms achieved the goal of Delhi's graded response action plan for effective mitigation strategies.

3.3.2.2 Data Integration and Preprocessing Framework

Three years of hourly data from five points in Delhi were monitored along with the Sentinel-5P tropospheric column to address the uneven spread of city pollution. These data can be prepared through various search patterns and min–max adjustments to detect both immediate changes and extended patterns in the NO_2 levels. This preparation system guarantees data integrity and uniformity, which is crucial for strong model effectiveness.

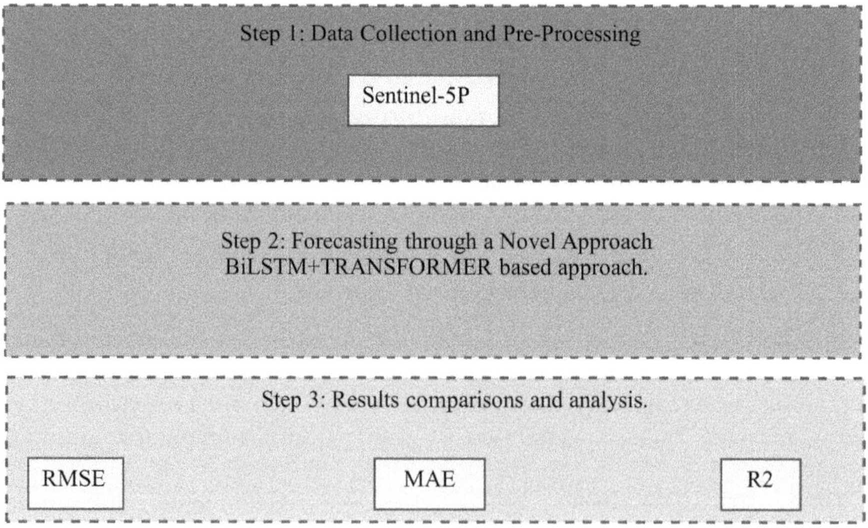

FIGURE 3.2
Overall framework of Breath-Net.

3.3.2.3 Hybrid Architecture Design and Implementation

The BREATH-Net framework features a hybrid architecture, which is depicted in Figure 3.3, including a Bi-LSTM model with transformers to identify time-related trends for understanding weather patterns and air quality levels. The transformer module contains four groups, where each group has a key size of 32 to adjust accordingly during ongoing forecasts with a dropout layer of 0.1, which avoids overfitting and provides a steady solution through normalization.

3.3.2.4 Training Methodology and Optimization

To implement the BREATH-Net framework, we set the learning rate to 0.001 using the Adam optimizer with 50 epochs. Early stopping can be used for 10 successive epochs to end the training once the improvement ceases. The dataset can be split at ratios of 0.8, 0.1, and 0.1 for model learning and for predicting the loss MSE used.

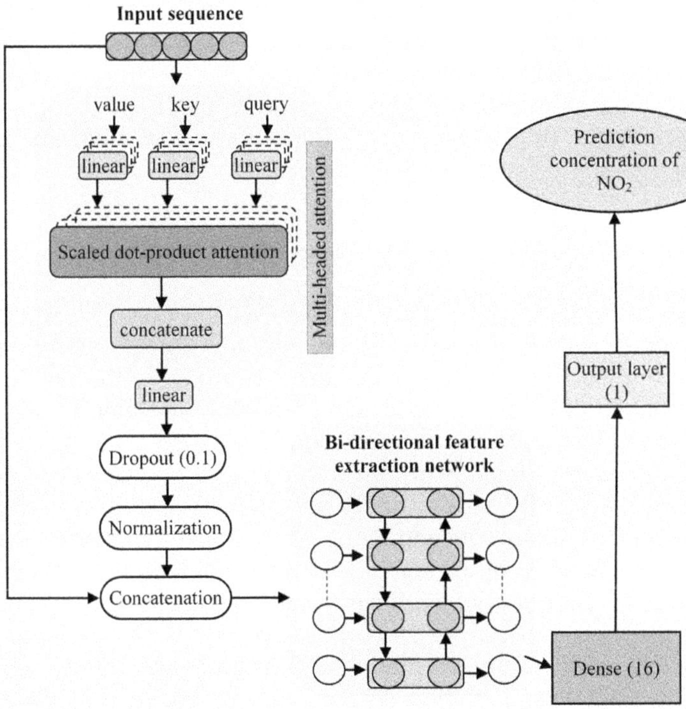

FIGURE 3.3
Architecture of multihead attention and BiLSTM for NO2 forecasting.

3.3.2.5 Performance Analysis and Validation Results

For the test data, the BREATH-Net framework achieves outstanding results, with an RMSE of 9.06 μg/m³, an MAE of 5.11 μg/m³, and an R^2 of 0.96, with XGBoost obtaining an RMSE of 12. The model also shows that it provides the best result when the weather is more consistent just before the monsoon, whereas it provides poor results in winters because of low temperatures and more pollution sources. The model's ability to maintain top performance in different weather patterns demonstrates its strength and usefulness in real situations.

3.3.2.6 Policy Applications and Urban Air Quality Management

The BREATH-Net framework offers actionable information for the management of Delhi's air quality via trustworthy hourly forecasts that allow for proactive intervention measures. Robust early warning capacities underpin dynamic traffic management and industrial emission regulations. The integration of the framework with Delhi's graded response action plan allows for evidence-based implementation of policies during pollution events. The real-time prediction function allows authorities to prepare health advisories for sensitive groups, such as children, elderly individuals, and people with preexisting respiratory diseases. The capacity of the framework to forecast peaks in pollution with adequate lead time allows the implementation of short-term emission-reducing measures. These applications illustrate the real-world benefits of advanced AI systems for urban environmental management.

3.3.3 Case Study 2: CATALYST Framework for $PM_{2.5}$ Forecasting

3.3.3.1 Framework Overview and Technical Innovation

The Convolutional and Transformer Model for Air Quality Forecasting (CATALYST) framework has made a valuable contribution to the prediction of particulate matter through the novel integration of pretrained convolutional neural networks (CNNs) with transformer-based temporal modeling. This architecture was developed for geographic and temporal problems in $PM_{2.5}$ forecasting, especially the nonlinear relationships between meteorological variables and pollutant concentrations. The framework was developed to enable proper recognition of the spatial and temporal features inherent to air quality data. The computational efficiency of CATALYST stems from its design, which uses a single encoder and prediction layer with a linear architecture, effectively minimizing the model's complexity and retaining predictive accuracy. This design enables practical use in real-time monitoring systems that may have limited computational resources. Because CATALYST can process 48-hour input windows while predicting 24-hour forecasts, it is well positioned for operational air quality management.

3.3.3.2 Data Sources and Environmental Context

CATALYST utilizes various $PM_{2.5}$ datasets over time from November 2016 to March 2023, covering the five monitoring stations across Delhi, namely, DTU, NSUT, Anand Vihar, Pusa, and IGI-T3. A longer period captures seasonal differences, extreme pollution episodes, and long-term changes in variations needed to develop a good model. The additional meteorological parameters assigned, such as humidity, boundary layer height, and land-use indices, provide an environmental context at the input level.

The data preprocessing stage uses mean imputation for any missing values and min–max normalization of the dataset to ensure that the features are on the same scale (measurement scale). A sliding window technique is used to create overlapping input sequences produced over 48 h to benefit from multiple days of temporal data that may affect $PM_{2.5}$ concentrations. Positional encoding allows temporal order to be retained in the dataset, which is useful for sequential data modeling and sequence modeling.

3.3.3.3 Hybrid Convolutional Neural Network – Transformer Architecture

The CATALYST architecture employs a pretrained convolutional neural network (CNN) for initial feature extraction from time-series data converted to two-dimensional representations. This approach leverages the CNN's proven capability for spatial pattern recognition while adapting it to temporal sequence analysis. The CNN component effectively reduces dimensionality while preserving essential temporal characteristics. This process is illustrated in Figure 3.4.

The transformer encoder processes CNN-extracted features through multihead self-attention mechanisms that capture long-range dependencies across temporal sequences. This combination enables the model to simultaneously utilize local temporal patterns identified by the CNN and global relationships discovered through attention mechanisms. The final linear layer integrates these multiscale features to produce accurate $PM_{2.5}$ predictions.

3.3.3.4 Training Optimization and Validation Methodology

CATALYST training employs the Adam optimizer with a learning rate of 0.0005, which is selected to balance convergence speed with stability. The Huber loss function provides robustness against outliers common in environmental datasets. The learning rate annealing schedule reduces rates by a factor of ten every 50 epochs, ensuring fine-tuned optimization in later training stages.

Model checkpoints based on validation MAEs enable the selection of optimal model configurations while preventing overfitting. The 80-10-10 dataset partition ensures comprehensive evaluation across training, validation, and testing phases. A batch size of 64 provides efficient gradient computation while maintaining sufficient stochasticity for robust optimization.

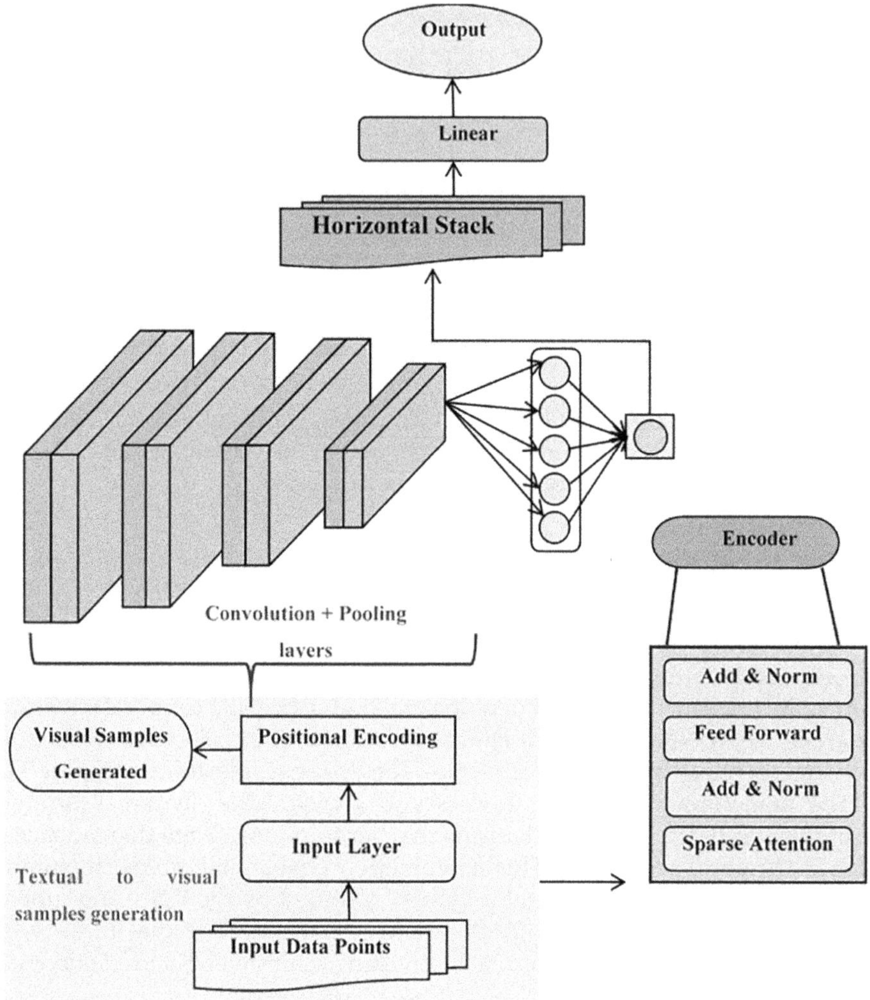

FIGURE 3.4
Architecture of CATALYST, a novel convolutional and transformer model for air quality forecasting.

3.3.3.5 Performance Evaluation and Cross-City Validation

CATALYST yields outstanding performance statistics, with an RMSE of 21.01 μg/m³, an MAE of 13.37 μg/m³, and an R^2 of 0.83 for the Delhi test data. Comparative analysis reveals better performance than independent LSTM models (RMSE: 25.71 μg/m³) and transformer-based models alone (RMSE: 24.02 μg/m³). These statistics highlight the efficacy of the hybrid approach in modeling intricate $PM_{2.5}$ dynamics. Cross-city verification tests in Mumbai and London show model generalizability, with RMSE increments of less than 20%. This transferability shows that the learned representations capture underlying meteorological condition–particulate matter relationships. The robust performance of the model across different urban domains indicates its wide applicability for monitoring air quality globally.

3.3.3.6 Operational Implementation and Environmental Impact

With the 24-hour $PM_{2.5}$ forecast, CATALYST supports decision-making for comprehensive air quality management options that include the control of industrial emissions to safeguard public health. The stock has been renowned for its accuracy during agricultural burning seasons, which is thus useful for transboundary pollution events. When implemented in real time, it may therefore be utilized to respond to an event of predicted pollution. Integration of this tool with monitoring networks can then enable automated production of alerts when forecast concentrations cross certain threshold levels on a health basis. These warnings will help to protect vulnerable populations and suggest protective measures.

3.4 Artificial Intelligence Architectures and Methodological Frameworks for Environmental Applications

3.4.1 Advanced Neural Network Architectures in Environmental Science

There has been great progress in the development of AI architectures for environmental evaluation by adopting transformer-based models that utilize self-attention mechanisms to capture long-range temporal dependencies, which allows for the joint extraction of multiple patterns within environmental data, such as recognizing changes in atmospheric conditions, oceanographic conditions, and terrestrial measurements all concurrently and outperforming recurrent neural networks. Physical architectures take advantage of the pattern recognition strengths of DL with respect to context, interpretations, and physical consistency provided by traditional environmental models, which incorporate solving inverse problems to improve the source identification or source location of pollution sources with high accuracy under complex atmospheric conditions without being overly computationally intensive. Recent implementations of multimodal spatiotemporal fusion models that combine CNNs with long short-term memory have improved the performance of air quality prediction.

3.4.2 Real-Time Processing and Edge Computing Implementation

The convergence of Internet of Things technology and edge AI has revolutionized real-time environmental monitoring capabilities on distributed sensor networks, reducing latency and bandwidth requirements while enhancing system resilience and supporting immediate anomaly detection and event-driven alarms, which is particularly valuable in scenarios where continuous internet connectivity cannot be guaranteed.

Modern edge computing implementations utilize microcontrollers and specialized processors for sensor interfacing and local AI processing, creating scalable and cost-effective monitoring networks. The integration of secure communication protocols ensures reliable data transmission to cloud-based visualization platforms while supporting real-time dashboards and automated alert systems. This distributed architecture enables environmental monitoring at unprecedented scales and temporal resolutions.

3.5 Emerging Technologies and Innovation Frontiers

3.5.1 Foundation Models for Earth System Monitoring

A foundation model is a potentially transformative technology that can be fine-tuned with small data requirements for different environmental applications, such as monitoring deforestation or predicting extreme weather events. Recent advances have focused on creating benchmarks for climate change applications to enable standardized evaluation of model performance across a large number of environmental tasks, foster collaborative design, and guarantee that foundation models meet the specific requirements of environmental applications while also adhering to standards of precision and dependability.

3.5.2 Digital Twin Technologies for Environmental Systems

Digital twins of Earth systems represent virtual replicas of environmental components that span multiple scales and domains, addressing complex environmental challenges from diverse sources, including satellite RS, IoT sensors, and traditional monitoring stations. This creates dynamic, interactive representations of environmental systems. Its key components include conceptual frameworks for standardized communication interfaces, scalable computational architectures, and adaptive management strategies for environmental decision-making.

3.5.3 Human–Environment Integration in Digital Systems

Modern digital twin implementations address both human impacts on environmental systems and human responses to environmental changes by incorporating social and economic factors alongside physical environmental processes, evaluation, and intervention analysis [10] to empower decision-makers to reduce risks and improve outcomes

3.6 Governance, Ethics, and Data Security in Artificial Intelligence Environmental Systems

3.6.1 Artificial Intelligence Governance Frameworks for Environmental Applications

The deployment of AI systems in environmental monitoring requires comprehensive governance frameworks that address data privacy, such as endangered species locations, model transparency, and decision accountability. Organizations must implement comprehensive security protocols across multiple operational domains, including data sourcing, model validation, and content verification, to ensure that model predictions and recommendations remain interpretable and accountable.

3.6.2 Federated Learning for Privacy-Preserving Environmental Monitoring

The federated learning paradigms can be very promising for collaborative environmental monitoring purposes, where for instance, AI models can be trained jointly without any sharing of raw environmental data by respecting data governance requirements and institutional policies. On the other hand, privacy preservation techniques such as differential privacy and secure multiparty computations keep the data secure while the actual collaborative training of models takes place.

3.6.3 Explainable Artificial Intelligence for Environmental Decision-Making

The development of explainable AI frameworks for model interpretability and decision transparency to understand the reasoning behind AI-generated environmental predictions and recommendations supports evidence-based policy development. Structural equation modeling combined with explainable AI approaches provides powerful tools for causal analysis to identify and quantify the relationships between different environmental factors of saliency maps and attention visualization techniques to influence model predictions.

3.7 Future Research Directions and Emerging Opportunities

3.7.1 Advanced Artificial Intelligence–Climate Modeling Integration

An analysis of historic climate data for finding patterns missed by traditional models in forecast prediction of weather events in real time and projection of long-term climate change effects. Better resource management and biodiversity monitoring capabilities are needed to address critical gaps in climate-related datasets.

3.7.2 Scaling and Implementation Challenges

Organizations must implement systematic approaches to overcome these challenges, with regular model evaluations held, stakeholder engagement, and iterative model improvements instigated to incorporate various data inputs and localized knowledge. Other feedback from users is also captured for the purpose of fine-tuning the model to ensure that the AI systems respond to changing requirements for environmental monitoring and to allow common assessments of AI system performance in different environmental applications.

3.7.3 Integration with the Circular Economy and Sustainability Principles

In the future, the development of scalable solutions that can support different geographical regions and environmental contexts but still maintain accuracy and reliability at the highest level is necessary. Green AI initiatives involve energy-efficient algorithm development, model architecture optimization, and the use of renewable energy sources for computing purposes. These multilayered sustainable resource optimization frameworks can greatly increase resource recovery efficiency and reduce transportation-related emissions via logistics optimization.

3.8 Conclusion

This section provides a conclusion on how hybrid AI architectures help real-world environmental challenges. The BREATH-Net and CATALYST frameworks demonstrate the ability of sophisticated neural networks to outperform all the pollution prediction algorithms studied here. The frameworks and innovations suggest a springboard into the advancement of planetary ecosystems for generations to come, and explainable AI holds the promise of enhancing our capacity to understand and protect the natural environment.

Acknowledgement

The authors express their gratitude to the reviewers and the editor for dedicating their valuable time to enhancing this work.

References

Fan, J., Qi, L., Hou, J., Feng, X., Karimian, H., & Lin, S. (2017). A Spatiotemporal Prediction Framework for Air Pollution Based on Deep RNN. *ISPRS Annals of the Photogrammetry, Remote Sensing and Spatial Information Sciences, 4,* 15–22. https://doi.org/10.5194/isprs-annals-iv-4-w2-15-2017

He, S., Dong, H., Zhang, Z., & Yuan, Y. (2022). An Ensemble Model-Based estimation of nitrogen dioxide in a southeastern coastal region of China. *Remote Sensing,* 14(12), 2807. https://doi.org/10.3390/rs14122807

Huang, C., Sun, K., Hu, J., Xue, T., Xu, H., & Wang, M. (2022). Estimating 2013–2019 NO2 Exposure with High Spatiotemporal Resolution in China using an Ensemble Model. *Environmental Pollution, 292,* 118285. https://doi.org/10.1016/j.envpol.2021.118285

Jia, X., Gong, X., Liu, X., Zhao, X., Meng, H., Dong, Q., Liu, G., & Gao, H. (2023). Deep Sequence Learning for Prediction of Daily NO2 Concentration in Coastal Cities of Northern China. *Atmosphere, 14*(3), 3. https://doi.org/10.3390/atmos14030467

Kumar, A., Shirin, S., Ansari, M. I., Pandey, G., Sharma, S. N., & Yadav, A. K. (2022). Fuzzy and neural network model-based environmental quality monitoring system: Past, present, and future. In S. P. Maurya, A. K. Yadav, & R. Singh (Eds.), *Modeling and Simulation of Environmental Systems: A Computation Approach,* 153–176. CRC Press. https://doi.org/10.1201/9781003203445-10

Pak, U., Ma, J., Ryu, U., Ryom, K., Juhyok, U., Pak, K., & Pak, C. (2020). Deep Learning-Based PM2.5 Prediction Considering the Spatiotemporal Correlations: A Case Study of Beijing, China. *Science of The Total Environment, 699,* 133561. https://doi.org/10.1016/j.scitotenv.2019.07.367

Patel, D., Kulwant, M., Shirin, S., Kumar, A., Ansari, M. A., & Yadav, A. K. (2022). Artificial Intelligence for Air Quality and Control Systems: Status and Future Trends. In S. P. Maurya, A. K. Yadav, & R. Singh (Eds.), *Modeling and Simulation of Environmental Systems: A Computation Approach,* 133–152. CRC Press. https://doi.org/10.1201/9781003203445-9

Rai, V., Kumar, S., Singh, T., & Kapoor, R. (2023). PM2.5 Level Forecasting using Transformer-Based Model. In *2023 3rd International Conference on Advance Computing and Innovative Technologies in Engineering (ICACITE),* 216–221. https://doi.org/10.1109/ICACITE57410.2023.10182822

Rastogi, U., Kumar, S., Maurya, S. P., & Yadav, A. K. (2025). Air Pollution Forecasting using Genetic Algorithm and LSTM Deep Learning Technique. *Aerosol Science and Engineering.* 10.1007/s41810-025-00343-y.

Sonata, I., & Heryadi, Y. (2024). Comparison of LSTM and Transformer for Time Series Data Forecasting. In *2024 7th International Conference on Informatics and Computational Sciences (ICICoS)*, 491–495. https://doi.org/10.1109/ICICoS62600.2024.10636892

Wang, Y.-Q., Wang, H.-C., Chen, J.-J., Yin, W.-X., Li, J., Zhang, Z., Li, H.-W., Wang, M.-F., & Wang, A.-J. (2025). Transformer Networks and Loss with Punishment for Optimized Management of Urban Water Supply System. *ACS ES&T Water*, *5*(2), 800–815. https://doi.org/10.1021/acsestwater.4c00881

4

Role and Applications of Remote Sensing in Climate Change and Detection: Hydro-Climatic Assessment and Projected Groundwater Recharge, Case Study

Shweta Kodihal, M.P. Akhtar, Satya Prakash Maurya, and Qipei Mei

4.1 Introduction

Remote sensing (RS) plays an essential role in detecting and monitoring climate change by acquiring data without physical contact with the target area. Typically, presented as digital grids or raster grids, RS data are invaluable for analyzing climate trends, validating climate models, and understanding the impact of climate change on ecosystems. Frequent satellite passes allow real-time observations of climate-related changes. Therefore, detection of climate change requires long-term data. For example, Wu et al. (2017) utilized time-series Landsat images to study the status of estuarine and coastal wetland ecosystems. Similarly, Kazemi Garajeh et al. (2024) analyzed key climatic variables such as air temperature, actual evapotranspiration, and total precipitation via the Moderate Resolution Imaging Spectroradiometer (MODIS) to monitor surface water resources under climate change. When incorporated with geographic information systems (GIS), RS techniques are crucial for analyzing spatially explicit and time-sensitive data (Sunkur et al., 2024; Rastogi et al., 2025). The combined use of RS and GIS tools is particularly effective for monitoring dynamic spatiotemporal changes in land use and land cover (Youssef et al., 2024; Kumar et al., 2022). Landsat imagery has been widely applied to create land use/land change maps, with RS and GIS providing excellent tools for evaluating spatial patterns (Baqa et al., 2021). In high-risk areas with limited resources, satellite RS offers a reliable alternative for climate change monitoring (Alvarez & Govind, 2024). A major challenge in regional-scale hydrological modeling is the need for highly accurate data. To address this, researchers have combined data from multiple satellites and models to produce high-resolution, precise datasets with wide spatial

DOI: 10.1201/9781042013234-5

coverage, thereby increasing the accuracy of precipitation datasets (Zhang et al., 2024). Statistical downscaling method (SDSM) software was used to downscale the precipitation and temperature data. Heidarzadeh et al. (2024) combined forecasted data with satellite imagery to estimate the future extent of wetland areas.

The ability of RS to monitor large-scale environmental changes makes it an indispensable tool for understanding and addressing climate change. Its integration with GIS enhances the spatial and temporal analysis of land-use patterns, making it essential for sustainable resource management and environmental protection efforts. As the impacts of climate change intensify, continued refinement of RS technologies will be the key to effective climate mitigation and adaptation.

4.1.1 Background

RS offers a powerful tool for identifying groundwater recharge zones by providing detailed, high-resolution data on land use patterns and climate variables, such as precipitation. The integration of satellite imagery with geospatial analysis enables precise mapping of areas with the highest potential for groundwater replenishment, aiding sustainable water resource management. Alexander et al. (2024) presented a multicriteria method that uses spatial data and RS, where various factors that affect groundwater are assigned weights and overlaid to obtain several classes of groundwater potential. Groundwater recharge zones can be identified via decision support systems by employing various influencing maps produced via RS and GIS techniques (Saad et al., 2024). Groundwater recharge is influenced by many hydrological, geological and climatic factors. Multicriteria decision-making (MCDM) is the link between these factors and recharge zone identification. MCDM has been widely employed by researchers to evaluate groundwater recharge zones. Kodihal & Akhtar (2024) discussed all the factors, methods of MCDM, and their weight allocation techniques employed by researchers in the identification of groundwater recharge zones. MCDM techniques involve the selection of factors, assigning weights to these factors, and overlaying them to demarcate the desired zone.

Jaipur city, the capital of Rajasthan state in India, has a semiarid climate, with an average temperature of 35° C and a maximum temperature of 48°C. The reported mean annual rainfall according to the Central Groundwater Board is 593.0 mm. Groundwater is one of the primary sources of water, especially in the Jaipur rural area. The growing population of the city has led to an increase in land use and a decrease in groundwater levels. Many researchers have addressed the increase in water supply in the study area (Patil et al., 2023; Gaur et al., 2011; Patil & Gupta, 2023). However, the issues of changing land use and rainfall have rarely been addressed in these studies. To gain a more comprehensive understanding of the potential changes and developments anticipated in a specific year, the present study emphasizes the use of projected land use and rainfall maps for the year 2040, incorporating these as critical climatic and land use criteria. This study employs a combination of AHP and ordered weight averaging to identify groundwater potential zones. This approach is designed to generate a series of maps that address various levels of risk associated with the identification of these recharge zones. A novel approach for locating runoff recharge wells, which involves superimposing the stream order map onto the groundwater recharge potential zones, is adopted in this study. This method enhances the precision of identifying optimal well locations by integrating surface water flow patterns with recharge potential areas.

4.1.2 Structure of the Chapter

The chapter is organized into five sections: introduction, methodology, validation, results and discussion, and conclusion. The first section provides an overview of the study area and outlines the background and objectives of the study. The second section details the methodology adopted, with each step explained in dedicated subsections. This includes the statistical downscaling technique used for rainfall prediction, Markov chain analysis applied for LULC forecasting, integration of thematic maps via both ordered and factorial weight approaches, and the method followed for runoff computation. The third section focuses on the validation process in which groundwater modeling is employed to assess the reliability of the approach. The fourth section presents and discusses the results obtained from various stages of the study. The fifth section concludes the chapter by summarizing the key findings and their implications.

4.2 Methodology

The methodology of this study integrates RS, statistical analysis, geospatial techniques, and numerical groundwater modeling to delineate and validate potential groundwater recharge zones under projected climate and land-use scenarios. The approach begins with the statistical downscaling of future rainfall data via the statistical downscaling model (SDSM), which incorporates projections from two shared socioeconomic pathway (SSP2-4.5 and SSP5-8.5) climate scenarios. These downscaled precipitation datasets served as critical inputs for subsequent hydrological analyses. Land use/land cover (LULC) changes are modeled via the Cellular Automata–Markov (CA–Markov) chain approach, with Landsat imagery from 2002 and 2012 analyzed to predict LULC distributions for the year 2022, capturing spatial development patterns and land transformation trends.

A set of nine thematic layers, namely, predicted rainfall, predicted LULC, aquifer thickness, aquifer type, transmissivity, lineament density, drainage density, soil type, and slope, are generated to evaluate recharge potential. These layers are integrated via ordered weighted averaging, with weights assigned through a combination of the analytical hierarchy process (AHP), expert judgment, and insights from the relevant literature. Runoff estimation is conducted using the Soil Conservation Service Curve Number (SCS-CN) method to complement the recharge assessment. The methodology culminates in the validation of identified recharge zones via a transient groundwater flow model developed in MODFLOW via the ModelMuse interface. Model performance was assessed by comparing simulated and observed groundwater heads, ensuring the robustness and reliability of the results. Each methodological component is described in detail in the subsequent sections.

4.2.1 Downscaling Precipitation

Downscaling of precipitation was conducted to predict the pattern and intensity of rainfall for the upcoming years. The future prediction of rainfall was performed by considering the changing climate scenarios provided by the Intergovernmental Panel on Climate Change (IPCC). The IPCC provides global climate models (GCMs) under its Coupled Model Inter-comparison Project (CMIP). Recently, a sixth assessment report on CMIP, including new shared socioeconomic pathways (SSPs), was released. These SSPs are SSP1--2.6, SSP2--4.5, SSP4--6.0, and SSP5--8.5, which provide five distinct narratives elucidating the overarching socioeconomic patterns exerting influence over the trajectory of future societies.

4.2.2 Statistical Downscaling Model

The downscaling of precipitation requires two sets of data: a predictor and predictand. The predictor is a set of daily precipitation data obtained from the Indian Meteorological Department. Predictors are obtained from the Copernicus Climate Change Service www.climate-scenarios.canada.ca/?page=main for the Norwegian Earth System Model (NorESM). NorESM has made a significant contribution to simulating climate, especially its application to studies evaluated in the IPCC's 5th assessment report (Seland et al., 2020). Establishing quantifiable correlations between local surface conditions (predictands) and large-scale atmospheric variables (predictors) is known as statistical downscaling. The downscaling model can be expressed as

$$R_t = F(X_T) \text{ for } T \leq t$$

Here, R_t is the local-scale predictand at time t; X_T is the set of predictors up to time t; and F is the technique used in quantifying the relationship between the predictand and predictor (Charles et al., 2004). The predictor sets derived for the study included the wind speed, zonal wind component, meridional wind component, relative vorticity, wind direction, true wind divergence for 1000 hPa, 500 hPa, and 850 hPa, mean sea level pressure, total precipitation, specific humidity, and air temperature at 2 m.

The National Centre for Environmental Prediction (NCEP) NCEP reanalysis data for the period 1979--2014 were used for model calibration. Predictors were selected via correlation analysis. This process involves examining partial correlation coefficients (partial *r*) and their associated *p*-values. Predictors are chosen based on partial *r* values that show the strongest correlations, with a focus on those with *p*-values near zero. The model was calibrated via the defined predictand and set of predictors to calculate multiple linear regression equations using the standard method. Several ensembles were used to calculate the average monthly precipitation, which was then cross-validated against the observed data. The correlation coefficient across all stations is between 0.7 and 0.9, which indicates a strong association. An RMSE value close to zero indicated a perfect fit. The Nash Sutcliff efficiency (NSE) acceptance criterion has a value greater than 0.5.

The downscaling technique adopted in this study is the 'weather generator.' The weather generator simulates possible weather scenarios based on statistical relationships and historical patterns. It generates a sequence of daily data based on observed climatic data and NCEP reanalysis data, along with regression model weights from the calibrated model. It is represented as:

$$R_t = F_w(\theta | X_T) \text{ for } T \leq t$$

where θ is the parameter set of the weather generator. Furthermore, a scenario generator was employed to downscale the NorESM predictors for the period 2015--2100 for SSP2 4.5 and SSP5 8.5. Rainfall values across the study area were obtained by interpolating the downscaled rainfall station values. Rainfall maps are prepared for the years 2030, 2040, and 2050 (Kodihal et al., 2024). However, in the present study, a downscaled rainfall map for only the year 2040 was employed, as it specifically concentrates on the scenario projected for that year.

4.2.3 Land Use Land Change

The Landsat series is highly regarded for its high resolution and consistent revisit times, making it an excellent choice for studying urbanization at various scales, including local, national, and international levels. For the temporal growth analysis of the study area, Landsat images with a resolution of 30m were utilized. This resolution allows for detailed observations of land cover changes over time, which is essential for understanding urban development patterns. Landsat images (Landsat 8-9 Operational Land Imager (OLI) and Thermal Infrared (TIRS) Collection 2 Level-2) for 2002, 2012, and 2022 were selected and analyzed across multiple spectral bands. The images were classified into five distinct land cover classes: built-up areas, vegetation, open lands, rangelands, and water bodies. This classification enables the quantification of changes in land use and the assessment of the impact of urbanization on the environment.

4.2.4 Land Change Model

Using the Terr Set software, a commercial GIS package, changes in land cover were meticulously examined from 2002 to 2012. This software provides advanced tools for spatial analysis and modeling, allowing for a comprehensive evaluation of urban growth and its implications. The primary objective of urban development is to forecast changes by attempting to represent the impacts of different growth drivers on land use and land change classes. These factors are typically categorized as temporal, spatial, or spatiotemporal (Yadav et al., 2019). Terrset's integrated software, the land change modeler (LCM), simplifies the complexity of the change analysis. In LCM, assessing and predicting land cover changes and their impacts are structured into change analysis, transition potential, and change prediction. Land cover change was analyzed between two time periods, land cover from 2002 and 2012. In the case of multiple land-cover types, the number of possible transitions is extensive. The Figure 4.1 shows the gains and losses for

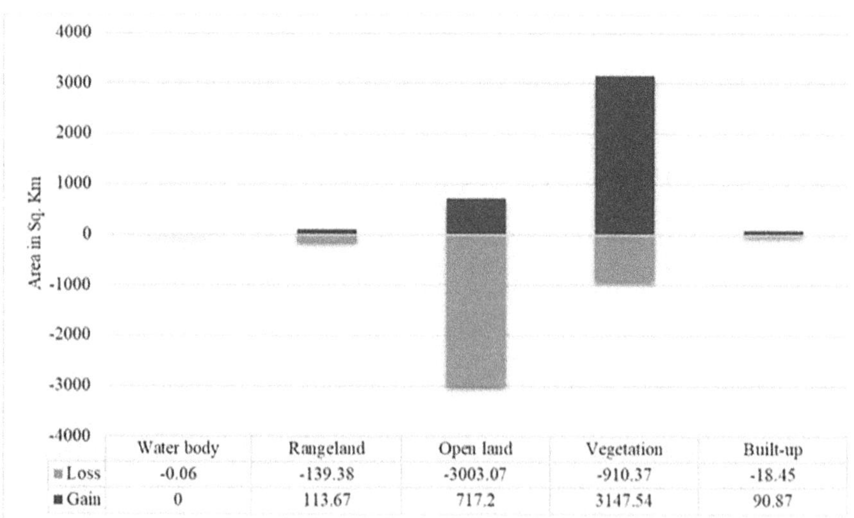

FIGURE 4.1
Land cover changes from 2002–2012.

each land cover class between the two time periods. The gains are shown on the positive axis and the losses are shown on the negative axis. The water body shows no significant loss, but shows a small gain. A significant loss was observed in the open-ground class with no gain. The built-up class had the largest gain among all classes, with no significant loss.

A transition potential model was employed to identify areas likely to undergo land transitions. The transition potential map shows the suitability for each transition created and organized into sub-models on the basis of the driver variables. The driver variables used in the study are distance to roads map, digital elevation model (DEM), slope, and urban-only map. All these driver variables are trained along with a sub-model that derives transitions from vegetation and open ground land classes. The selected sub-models are modeled via a multilayer perceptron (MLP). The MLP models pixel band values as input neurons and land classes as output neurons. The hidden layers between the input and output are referred to as perceptrons (ahurley, 1987). The model training was finalized after several iterations when the error was within the specified range. When the precision approaches the maximum number of iterations, transient potential maps are generated for each sub-model, showing the appropriate fit of each LULC class for the predicted year 2022. The LULC map for the year 2022 was modeled in the change prediction tab via Markov chain analysis. The Markov model uses a transition probability matrix to represent the changes in each thematic class during a specified time period. Memarian et al. (2012) and Alqadhi et al. (2021) described the Markov chain as follows:

$$\begin{aligned} F_x\left(X\left(T_{m+1}\right) \leq \left\{x_{m+1} \middle| X\left(T_m\right)\right\} = x_m, X(T_{m-1}) = x_{m-1}, \ldots, X(T_1) = x_1\right) \\ = (X(T_{m+1}) \leq (x_{m+1} | X(T_m))) = x_m \end{aligned} \tag{4.1}$$

Here, $X(T)$ signifies the Markov chain at time (t), (T_m) defines the current time, (T_{m-1}) represents the past time, and (T_{m+1}) defines the future time. The probability of transitioning from state i to state j is given by

$$P_{i,j} = P_r(X[m+1] = \left\{j \middle| X[m]\right\} = i) \tag{4.2}$$

Here, X[m] denotes the state {x1, x2, x3,... .} for (i) and (j), respectively. The transition probability matrix is given by (Memarian et al., 2012; Alqadhi et al., 2021)

$$\begin{bmatrix} P_{1,1} & P_{1,2} & . & . & P_{1,n} \\ P_{2,1} & P_{2,2} & . & . & P_{2,n} \\ . & . & . & . & . \\ . & . & . & . & . \\ P_{n,1} & P_{n,2} & . & . & P_{n,n} \end{bmatrix} \tag{4.3}$$

The Markov chain plots the stochastic process using Eq. (4):

$$\prod(T+1) = p^n x \prod(T) \tag{4.4}$$

Here, $\prod(T)$ is the system state at time t, $\prod(T+1)$ is the system state after interval (T+1), and p^n represents the potential state.

The performance of the modeled LULC map for 2022 was evaluated using the original 2022 map prepared via Landsat. The comparison revealed an approximate match of 78%, indicating a reasonable level of accuracy in the model's prediction, as shown in Figure 4.2. A scenario indicating both sub-models was employed further to project LULC maps for the years 2030, 2040, and 2050 (Kodihal & Akhtar, 2023).

4.2.5 Overlaying of Factor Maps

Nine factors that influence the recharge potential of groundwater in the study area were identified. The nine factors are rainfall, land use/land change (LULC), aquifer thickness, aquifer type, transmissivity, lineament density, drainage density, soil type, and slope, as presented in Figure 4.3. These factors include hydrological, geological, topographical, and climatic factors. To account for potential future changes, projected data for rainfall and LULC for the year 2040 were considered. All factorial maps were reclassified into five suitability classes, as mentioned in Table 4.1.

4.2.6 Ordered Weighted Averaging

In a spatial multi-criteria decision problem, a collection of viable options is evaluated via several criteria, many of which are incompatible and challenging to compare. (Boroushaki & Malczewski, 2008). The Boolean overlay and weighted summation procedures can be generalized using the ordered weighted averaging (OWA) procedure. The multicriteria aggregation method known as OWA was created within the framework of fuzzy set theory (Yager, 1988). OWA uses two sets of weights: order weights and criterion significance weights. A wide range of maps can be generated by adjusting the order weights. Several approaches for the application of OWAs have been reported (Malczewski, 1999; Malczewski & Liu, 2014; Malczewski et al., 2003; Makropoulos & Butler, 2006; Mokarram et al., 2021; Rahman & Szabó, 2022).

4.2.7 Analytical Hierarchical Process

Criteria importance weights can be assigned via the analytical hierarchical process (AHP). This method was introduced by Saaty (1977), and involves the use of a pairwise comparison matrix. AHP, which is an extension of OWA, was employed in this study to address the spatial decision problem.

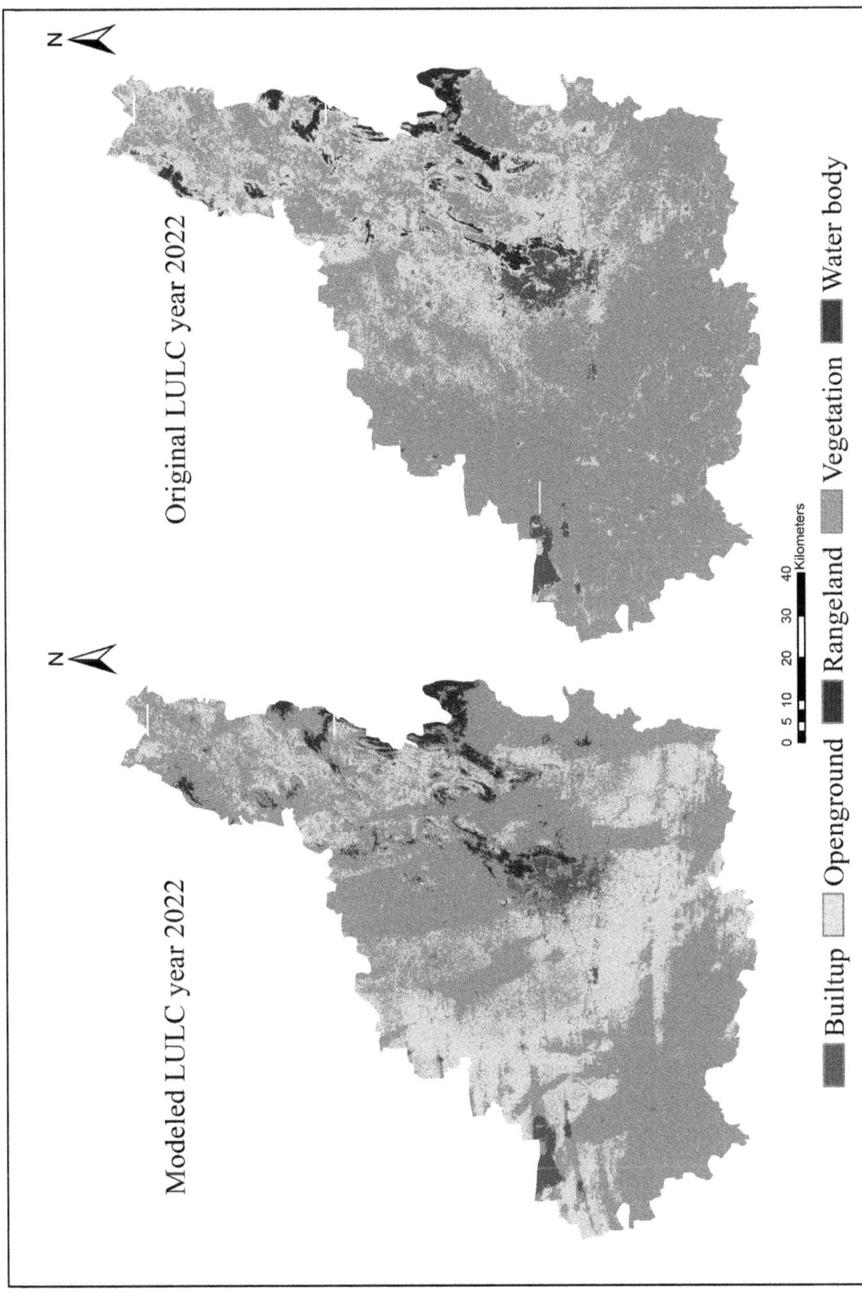

FIGURE 4.2
Modeled and original land -use/land cover map of 2022.

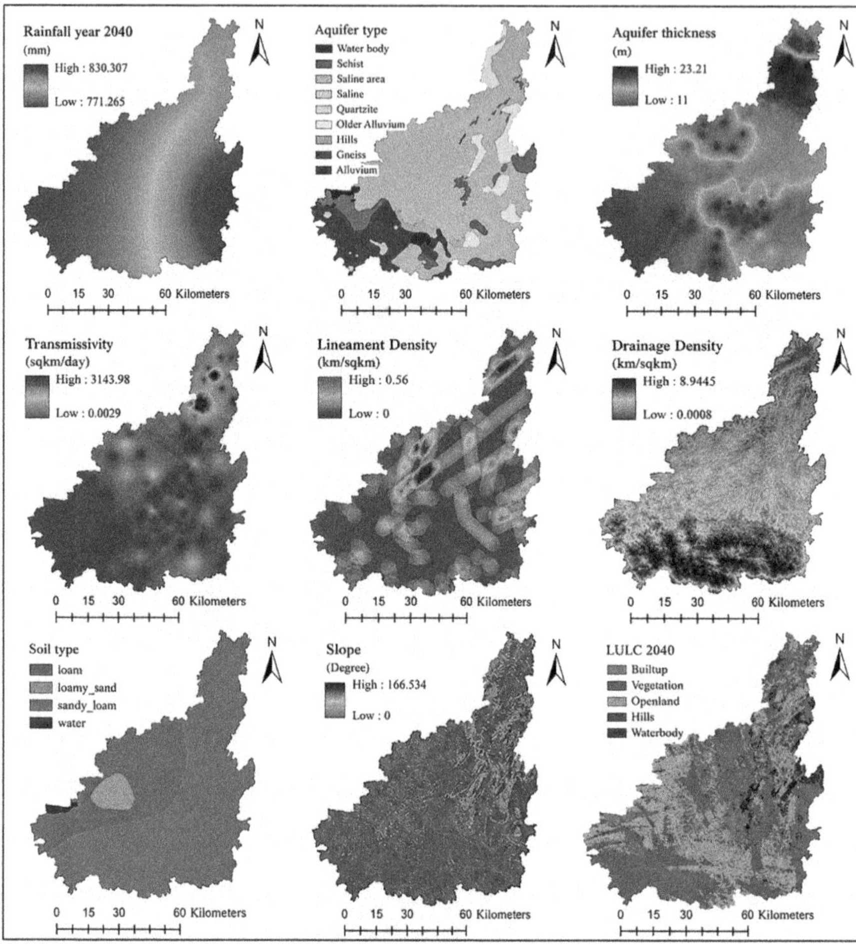

FIGURE 4.3
Generated maps of transmissivity, aquifer, aquifer thickness, lineament density, drainage density, soil type and slope for the study area along with projected rainfall and land-use/land cover maps for the year 2040.

AHP was applied in four steps: (1) carefully selecting the factors influencing groundwater recharge zones, (2) creating a pairwise comparison matrix with a nine-point rating scale (Table 4.2), (3) estimating relative weights, and (4) evaluating matrix consistency (Lentswe & Molwalefhe, 2020). This pairwise comparison matrix was developed to identify recharge potential zones based on the number of input factors (Saaty, 1989).

The pairwise comparison matrix is represented by $B = [b_{kl}]_{p \times q}$, where b_{kl} denotes the rating of the pairwise comparison matrix between criteria k and l. Matrix B is a reciprocal matrix, meaning $b_{kl} = \frac{1}{b_{kl}}$, and all diagonal elements are

TABLE 4.1

Reclassification of factors into five categories

S. No	Factors	Classes	Ranks
1	Rainfall (mm)	289.21-303.53	5
		303.53–315.76	4
		315.76–328.29	3
		328.29–341.72	2
		341.72–364.99	1
2	Transmissivity (m^2/day)	0.0030–73.98	5
		73.98–517.83	4
		517.83–1109.64	3
		1109.64–2206.95	2
		2206.95–3143.97	1
3	Aquifers	Saline area	5
		waterbody	5
		gneiss	4
		schist	3
		hills	3
		quartzite	2
		Older alluvium	1
4	Aquifer thickness (m)	11.00–14.97	5
		14.97–16.98	4
		16.98–18.80	3
		18.80–20.62	2
		20.62–23.20	1
5	Lineament density (km/sq.km)	0–0.035	5
		0.035–0.11	4
		0.11–0.19	3
		0.19–0.32	2
		0.32–0.56	1
6	Drainage density (km/sq.km)	0.00076–1.298478	5
		1.298478–2.771562	4
		2.771562–4.314794	3
		4.314794–5.963246	2
		5.963246–8.944489	1
7	Slope (degree)	56.164515–166.534317	5
		34. 6130–56.1645	4
		16.326894–34.613015	3
		4.5715–16.3268	2
		0-4.5715	1
8	LULC (land use classes)	Built-up	5
		Waterbody	4
		Hills	3
		Open-ground	2
		Vegetation	1
9	Soil Type	Loamy-sand	2
		Loam	2
		Sandy-loam	1

TABLE 4.2
Nine-point scale

1/9	Extremely less important
1/7	Very Strongly less important
1/5	Strongly less important
1/3	Moderately less important
1	Equally important
3	Extremely more important
5	Very Strongly more important
7	Strongly more important
9	Moderately more important

equal to one, i.e., $b_{kl} = 1$ for $k = l$. A similar principle applies at the attribute level. At this level, for each criterion, a pairwise comparison matrix was constructed by comparing the respective attributes. Thus, $B(k) = [c_{mn}(i)]_{r\times r}$ for $i = 1,2, \ldots, p$, where $c_{mn}(i)$ is the rating of pairwise comparison between attribute m and attribute n associated with criterion i (Boroushaki & Malczewski, 2008).

To calculate the weight of each criterion, the elements in each column of the pairwise comparison matrix are summed, as presented in Table 4.3. Each element in the matrix is divided by the sum of its respective columns via equation (4.5). The average of the values in each row of the normalized matrix is subsequently calculated using eqs. 4.5 and 4.6 (Mokarram et al., 2021).

$$r_{kl} = \frac{b_{kl}}{\sum_{k=1}^{p} b_{kl}} \tag{4.5}$$

$$w_k = \frac{1}{n}\sum_{k=1}^{r} r_{kl} \tag{4.6}$$

Where p is the number of columns, r is the number of rows, b_{kl} is the pairwise comparison matrix, r_{kl} is the normalized value for the element and w_k is the weight for k. The final score is calculated using Equation (4.7).

$$V_x = \sum_{k=1}^{r} W_k(g_{kl}) \tag{4.7}$$

Where V_x is the final score of k, w_k is the weight for k, and g_{kl} is the rating corresponding to criterion k.

The consistency index (CI) was evaluated using Eq. 4.8, where λ_{max} is the principal eigenvalue and r is the number of columns in the matrix.

$\lambda_{max} = \sum_{k=1}^{r} \frac{PS_k}{w_k}/r$. Where $PS_k = \sum_{l=1}^{r} q_{kl}/r$, with $q_{kl} = b_{kl}/w_k$. (Gdoura et al., 2015)

$$CI = \frac{(\lambda_{max} - r)}{(r-1)} \tag{4.8}$$

TABLE 4.3

Normalized pairwise comparison matrix for the analytical hierarchy process considering nine key factors

Key factor	Rainfall	Aquifer	Aquifer thickness	Transmissivity	Lineament Density	Drainage density	Soil	Slope	LULC	Principal Eigen Vector (Weights)
Rainfall	1	2	3	4	5	5	6	7	8	0.3058
Aquifer	1/2	1	2	3	4	5	6	5	8	0.2259
Aquifer thickness	1/3	1/2	1	2	3	3	4	4	5	0.1437
Transmissivity	1/4	1/3	1/2	1	2	2	3	3	2	0.09
Lineament density	1/5	1/4	1/3	1/2	1	2	3	4	5	0.0815
Drainage density	1/5	1/5	1/3	1/2	1/2	1	2	3	4	0.0598
Soil	1/6	1/6	1/4	1/3	1/3	1/2	1	2	2	0.0381
Slope	1/7	1/5	1/4	1/3	1/4	1/3	1/2	1	2	0.0310
LULC	1/8	1/8	1/5	1/2	1/5	1/4	1/2	1/2	1	0.0243

Where CI is the

$$CR = \frac{CI}{RI} \tag{4.9}$$

CR is the consistency ratio, and RI is a random index that depends on the number of elements being compared.

4.2.8 Order Weight

Order weights are the second set of weights assigned to the factors to control how these factors are overlaid. The factor weight applies to all pixels of the factor map, meaning that each factor map receives the same factor weight. This weight determines the relative importance of each factor in identifying the groundwater recharge zone and influences how the factors are traded off against one another. A factor with a high weight can balance out low scores for other factors, even if its unsuitability score is not remarkably high. While factor weights dictate how the factors compensate for each other, order weights establish the overall extent of this compensation. Boolean approaches are extreme functions that yield either very risk-averse solutions when the AND operator is applied or risk-taking solutions when the OR operator is used (ahurley, 1987).

The application of order weights facilitates the aggregation of solutions that can be positioned anywhere along the risk continuum between the AND and OR approaches. Order weights are fundamentally distinct from factor weights, as they are not associated with any specific factor. Instead, they are implemented on a pixel-by-pixel basis, reflecting the rank ordering of the factor scores at each location (pixel). For each pixel, an order weight of 1 was assigned to the factor with the lowest rank (i.e., the factor with the lowest score), an order weight of 2 was assigned to the next highest-ranked factor, and so on. Consequently, a single-order weight may be assigned to pixels from various factors based on their relative rank order (ahurley, 1987).

The OWA framework provides a mechanism for balancing multiple criteria while adhering to predefined constraints. Figure 4.4 shows a risk-tradeoff graph, which displays the decision-making method with positions labeled (a), (b), (c), (d), (e), and (f). These positions represent different combinations of risk and tradeoff: 'average risk-no tradeoff' (ARNT), 'low risk-no tradeoff' (LRNT), 'high risk-no tradeoff' (HRNT), 'average risk-full tradeoff' (ARFT), 'low risk and average tradeoff' (LRAT), and 'high risk and average tradeoff' (HRAT).

In position (b), the operator exhibits complete AND-ness without compromise, as represented by [1,0,0…0]. Position (c), on the other hand, highlights the operator's ideal performance, where it reaches full OR-ness with no tradeoff, as depicted by [0...,1]. Position (a) reflects a balanced mix of both AND-ness and OR-ness, represented by the vector [0,..1..,0]. In position (d), the operator takes an averaging approach, demonstrating a middle ground between AND and OR, while also accounting for trade-offs through equal-order weights (Eastman, 1999). Position (e) increasingly favors AND-ness, with the weights gradually approaching zero as the matrix moves toward OR-ness. Conversely, position (f) emphasizes OR-ness, where the weights tend toward zero in the direction of AND-ness within the decision triangle (Kodihal & Akhtar, 2023).

The OWA framework is versatile, permitting any combination of weights that sums to 1.0 to be beyond the specified scenarios. Regardless of the specific assignment of order weights, the resulting decision rule is situated within a triangular decision strategy space, as depicted in Figure 4.4 (Ahurley, 1987). This flexibility allows for the determination of appropriate weightings based on the relative importance of various factors (Drobne & Lisec, 2009; Eastman, 1999; Eastman, 2012). For further insights into OWA applications combined with AHP, see the works of Malczewski et al. (2003), Mokarram et al. (2021), and Mussa et al. (2020). The order weights assigned to the ordered maps for this study are presented in Table 4.4

The reclassified map was further overlaid in OWA, considering the six strategies producing six resultant maps. The maps in the ARNT and ARFT categories mainly showed zones in classes 2 and 3, whereas the HRNT and HRAT categories highlighted zones in classes 4 and 5. The map under the 'low risk average tradeoff' (LRAT) category predominantly shows zones in

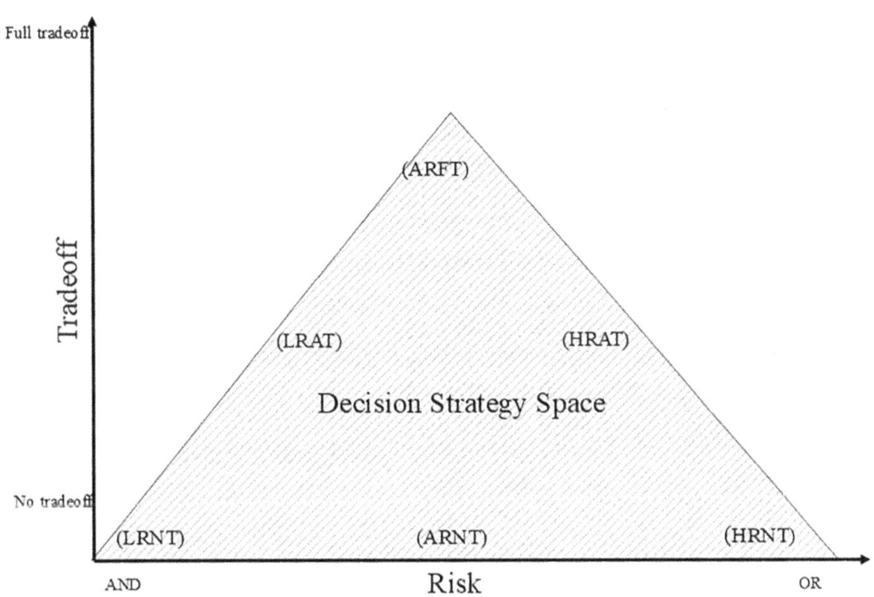

FIGURE 4.4
Revised decision strategy in ordered weighted averaging as recommended by (Eastman J R, 2012) and (Drobne & Lisec, 2009).

TABLE 4.4
Order weights for the various strategies applied in the present study

Decision Criteria	Order Weights								
	w_1	w_2	w_3	w_4	w_5	w_6	w_7	w_8	w_9
a) ARNT	0	0	0	0	1	0	0	0	0
b) LRNT	1	0	0	0	0	0	0	0	0
c) HRNT	0	0	0	0	0	0	0	0	1
d) ARFT	0.11	0.11	0.11	0.11	0.11	0.11	0.11	0.11	0.11
e) LRAT	0.4	0.28	0.12	0.06	0.05	0.04	0.03	0.02	0.00
f) HRAT	0.00	0.02	0.03	0.04	0.05	0.06	0.12	0.28	0.4

class 1. Therefore, this map was employed to identify suitable locations for recharge wells.

4.2.9 Run-Off Calculations

The class 1 map of the study area was extracted and overlaid with a stream order map to specifically locate recharge wells at the confluences of higher-order streams (Kodihal & Akhtar, 2023). It is important to calculate the amount of water that can be injected into recharge wells to ensure the efficient and sustainable operation of the system. Accurate estimation prevents overloading of the well, which could lead to clogging or reduced recharge capacity, while also optimizing the use of available runoff to increase groundwater replenishment. The runoff calculation was performed using the runoff curve number. The curve number (CN) method was developed by the Natural Resources Conservation Service (NRCS) of the United States Department of Agriculture (USDA), previously known as the Soil Conservation Service (SCS). This method is based on an empirical analysis of runoff patterns observed in small catchments and hillslope plots monitored by the USDA. It has proven to be highly effective in estimating the depth of direct runoff for specific rainfall events across various geographic regions. Mishra & Singh (2003) provided a comprehensive table of CN values based on different combinations of soil type, vegetation cover, and land use, which was utilized in this study. The CN method incorporates factors such as soil type, land use, and surface conditions to accurately predict runoff in watersheds. The curve number is related to the soil moisture retention (S) using the following equation:

$$S = \frac{25400}{CN} - 254 \tag{4.10}$$

where S is the potential maximum storage and CN is the curve number.

The runoff quantity is estimated using the following equation:

$$Q = \frac{(P - 0.2S)^2}{(P + 0.8S)} \tag{4.11}$$

where Q represents the runoff depth in millimeters, P denotes the precipitation in millimeters, and S refers to the potential maximum storage.

Patil et al. (2023) highlighted the application of the SCS-CN method to evaluate the rainwater harvesting potential in the extended urban area of Jaipur. The soil in the study area mainly consists of loam and sand, where sandy loam is classified as hydrological soil group (HSG) C, and loamy soil is classified as soil group D (Khare D. et al., 2017). The land-use map was prepared from Landsat 8-9 images from the United States Geological Survey (USGS) (https://earthexplorer.usgs.gov). The soil map is delineated for the study area from the Food and Agriculture Organization www.fao.org/soils-portal/. The curve number for the LULC 'vegetation' class on soil HSG C is 86, and the 'open ground' class on soil HSG D was 89. The number of urban settlements (built-up) in soil groups HSG C and HSG D was 98 (Kodihal et al., 2024). The surface runoff calculated for the year 2040 is 146.95 mm in depth, as presented in Table 4.5.

4.2.10 Modeling the Groundwater

Analyze the volumetric changes in groundwater resulting from the implementation of recharge wells by simulating groundwater flow patterns. The groundwater model was calibrated and validated in MODFLOW, which demonstrated high reliability with a correlation coefficient of 0.95. A comparison of the water budget before and after the application of recharge wells revealed an increase of 24% in water volume for the year 2040. This significant increase in groundwater volume underscores the effectiveness of recharge wells as a viable strategy for enhancing groundwater resources in the study area. The simulation results indicate that recharge wells not only improve the water table levels but also contribute to better water quality by facilitating the natural filtration process as water percolates through these soil layers.

TABLE 4.5

Curve number and surface run-off depth

	HSG C	HSG D	Runoff (mm in depth)
Built-up	98	98	146.95
Vegetation/open-ground	86	-	159.82
Vegetation/open-ground	-	89	164.14

The Modular-Finite Difference Groundwater Flow Model (MODFLOW), developed by the U.S. Geological Survey (USGS) is a software program designed to simulate typical features in groundwater systems. Created in the early 1980s, MODFLOW has continuously advanced with the integration of many new modules and related applications in groundwater studies. Currently, it is the most widely used program globally for simulating groundwater flow. As open-access software, it enables users to conduct tests and produce results. It employs a finite-difference method to describe groundwater flow at the grid level, making it ideal for modeling and simulating groundwater movement to evaluate groundwater resources. The MODFLOW model breaks down aquifers into grid cells, which can be organized into rows, columns, and layers.

$$\left[\frac{\partial}{\partial x}\left(K_x\frac{\partial h}{\partial x}\right)+\frac{\partial}{\partial x}\left(K_y\frac{\partial h}{\partial y}\right)+\frac{\partial}{\partial x}\left(K_z\frac{\partial h}{\partial z}\right)\right]-W=S_s\frac{\partial h}{\partial t}$$

where

Kx, Ky, and Kz are the hydraulic conductivities in the x, y, and z directions, respectively;

h represents the head height at position (x,y,z) and time t.

W denotes the volumetric flux per unit volume, which represents the source or sink.

Ss is specific storage.

4.3 Results and Discussions

The downscaled rainfall projections for the SSP2-4.5 and SSP5-8.5 scenarios highlight significant spatial and temporal variability across the study area for 2030, 2040, and 2050. While SSP2-4.5 generally predicts higher rainfall in earlier years, SSP5-8.5 shows an increase by 2050, with the northeast, middle, and southwestern zones experiencing shifting rainfall intensities. The runoff depth, computed via the SCS-CN method, reflects these variations, with the maximum runoff (485.2 mm) observed in vegetated/open lands under SSP2-4.5 in 2030 and the lowest runoff (94.5 mm) in 2040 under the same scenario. These results underscore the influence of both land cover and precipitation scenarios on surface runoff, informing site-specific recharge potential analyses. The Land Change Modeler in TerrSet, supported by MLP and CA–Markov analysis, successfully simulated future land-use patterns using Landsat data from 2002 and 2012. Validation against the actual 2022

LULC map yielded an accuracy of 78.56%, indicating reliable performance. The projected maps for 2030, 2040, and 2050 revealed a consistent increase in built-up areas, reflecting ongoing urban expansion. The year 2040 was selected for further analysis because of its representative balance of land-use shifts, making it a critical reference point for assessing groundwater recharge potential.

To delineate recharge zones, the study integrates rainfall projections from SSP4.5 scenarios and applies six combinations of risk and trade-off criteria to generate a total of six recharge zone maps. The most suitable recharge zones are identified under the 'good' category, particularly using the average-risk, full trade-off criteria. A total of 13 recharge wells were identified, with the majority situated in the eastern part of the study area. This spatial distribution aligns with the concentration of suitable recharge zones delineated under average-risk, full trade-off criteria, highlighting the region's greater potential for effective groundwater replenishment. The groundwater model developed in MODFLOW showed high reliability, with a correlation coefficient of 0.95. The simulation results revealed a 24% increase in groundwater volume for the year 2040, following the implementation of recharge wells. This demonstrates the effectiveness of recharge wells in enhancing water availability.

4.4 Conclusion

This study is important for addressing the challenges posed by rapid urbanization and climate variability, particularly in semiarid regions where groundwater serves as a primary water resource. By focusing on Jaipur, this study provides crucial insights into how RS and predictive modeling can be used to proactively manage groundwater resources. RS was effectively used in this study to project LULC changes in the study area, leveraging high-resolution satellite imagery for accurate and timely assessments. By employing Landsat images, this study identified significant shifts in urban and vegetation cover, which are instrumental in understanding future impacts on groundwater resources. This comprehensive projection is essential for adapting water management to ongoing urban expansion and climate change.

The study further integrates this projected rainfall and LULC data to delineate groundwater recharge potential zones, extending the applicability of the findings through 2040. By overlaying critical environmental factors, including soil type and drainage density, this study revealed areas with optimal conditions for groundwater replenishment, facilitating sustainable resource allocation amid increasing water demand.

Additionally, the RS data contributed to precise runoff calculations, enabling the quantification of the surface water available for recharge. Using the runoff curve number method, which is calibrated to regional soil and vegetation characteristics, this study provides accurate assessments of potential recharge inputs and optimizes groundwater recharge strategies.

Digital elevation models (DEMs) were also processed to generate a streamflow map, which was overlaid onto the recharge zones to identify ideal locations for recharge wells. This integration of DEM-based streamflow data allows for targeted well placement along high-order stream confluences, ensuring the maximum capture and infiltration of runoff.

Finally, groundwater modeling via MODFLOW successfully simulated the impact of recharge wells, revealing a significant increase in groundwater volume after application. This model demonstrates the effectiveness of strategically placed recharge wells and offers a reliable framework for enhancing the water availability and improving the water table in the study area.

The insights from RS inputs support spatial decision-making, enabling the implementation of recharge wells in zones most vulnerable to climate-induced water stress. Through this targeted approach, RS contributes to mitigating the adverse effects of climate change by increasing groundwater availability and promoting sustainable resource utilization.

Acknowledgement

The authors would like to thank the Central Groundwater Board, Department of Water Resources, River Development and Ganga Rejuvenation, Government of India, Western Region, Jhalana Doongri, Jaipur for providing well and aquifer data. Authors would like to acknowledge United States of Geological Survey (USGS) website https://earthexplorer.usgs.gov, Food and Agriculture Organization of the United Nations www.fao.org/soils-portal/ for providing the required data.

References

ahurley. (1987). *TerrSet manual.* www.forums.clarklabs.org/hc/en-us

Alexander, A. C., Sidiropoulos, P., Mohamed, E., El-Bana, M., Alogayell, H. M., Sheta, M. H., & Abdelfattah, M. (2024). An integrated remote sensing and GIS-based technique for mapping groundwater recharge zones: A case study of SW Riyadh, Central Saudi Arabia. *Hydrology, 11*(3), 38. https://doi.org/10.3390/HYDROLOGY11030038

Alqadhi, S., Mallick, J., Balha, A., Bindajam, A., Singh, C. K., & Hoa, P. V. (2021). Spatial and decadal prediction of land use/land cover using multilayer perceptron-neural network (MLP-NN) algorithm for a semiarid region of Asir, Saudi Arabia. *Earth Science Informatics, 14*(3), 1547–1562. https://doi.org/ 10.1007/s12145-021-00633-2

Alvarez, C. I., & Govind, A. (2024). Using remote sensing data in the cloud to monitor climate change in Senegal regions based on seasonal variables from 2000 to 2020. An Opportunity to Sustainable Policies, *IGARSS 2024 - 2024 IEEE International Geoscience and Remote Sensing Symposium*, Athens, Greece, 2024, 921–925. https://doi.org/10.1109/IGARSS53475.2024.10641322

Baqa, M. F., Chen, F., Lu, L., Qureshi, S., Tariq, A., Wang, S., Jing, L., Hamza, S., & Li, Q. (2021). Monitoring and modeling the patterns and trends of urban growth using urban sprawl matrix and CA-Markov model: A case study of Karachi, Pakistan. *Land, 10*(7), 700. https://doi.org/10.3390/land10070700

Boroushaki, S., & Malczewski, J. (2008). Implementing an extension of the analytical hierarchy process using ordered weighted averaging operators with fuzzy quantifiers in ArcGIS. *Computers & Geosciences, 34*(4), 399–410. https://doi.org/10.1016/J.CAGEO.2007.04.003

Charles, S. P., Bates, B. C., Smith, I. N., & Hughes, J. P. (2004). Statistical downscaling of daily precipitation from observed and modeled atmospheric fields. *Hydrological Processes, 18*(8), 1373–1394. https://doi.org/10.1002/HYP.1418

Drobne, S., & Lisec, A. (2009). Multi-Attribute Decision Analysis in GIS: Weighted Linear Combination and Ordered Weighted Averaging. *Informatica, 33*, 459–474.

Eastman, J. R. (1999). Multi-criteria evaluation and GIS. In: P. A., Goodchild, M. F., Maguire, D. J., & Rhind, D.W., Eds., *Geographical Information Systems, Longley*, John Wiley and Sons, New York, 493–502.

Eastman, J. R. (2012). *Idrisi Selva tutorial, manual version 17*. Clarks Labs. Clark University.

Gaur, S., Chahar, B. R., & Graillot, D. (2011). Combined use of groundwater modeling and potential zone analysis for management of groundwater. *International Journal of Applied Earth Observation and Geoinformation, 13*(1), 127–139. https://doi.org/10.1016/J.JAG.2010.09.001

Gdoura, K., Anane, M., & Jellali, S. (2015). Geospatial and AHP-multicriteria analyses to locate and rank suitable sites for groundwater recharge with reclaimed water. *Resources, Conservation and Recycling, 104*, 19–30. https://doi.org/10.1016/J.RESCONREC.2015.09.003

Heidarzadeh, N., Eghbal, M., Esmaeili, N., & Panaghi, K. (2024). Predicting the impact of climate change on the area of wetlands using remote sensing. *Theoretical and Applied Climatology, 155*(6), 5593–5606. https://doi.org/ 10.1007/s00704-024-04969-9

Kazemi Garajeh, M., Haji, F., Tohidfar, M., Sadeqi, A., Ahmadi, R., & Kariminejad, N. (2024). Spatiotemporal monitoring of climate change impacts on water resources using an integrated approach of remote sensing and Google Earth Engine. *Scientific Reports*, 14(1), 1–16. https://doi.org/10.1038/s41598-024-56160-9

Khare, D., Singh, L., & Jat, M. K. (2017). Impact of landuse change on urban runoff: A case study of Jaipur City, India. *Proceedings of the 37th IAHR World Congress*. www.researchgate.net/publication/319178751_Impact_of_Landuse_Change_on_Urban_Runoff_A_Case_Study_of_Jaipur_City_India

Kodihal, S., & Akhtar, M. P. (2024). Sustainable groundwater recharge potential zone identification: An AHP-OWA approach integrating future rainfall and land-use projections. *Water Resources Management Water Resources Management: An International Journal, Published for the European Water Resources Association (EWRA), Springer;European Water Resources Association (EWRA),, 38*(3), 1079–1098. https://doi.org/10.1007/S11269-023-03710-X/METRICS

Kodihal, S., & Akhtar, M. P. (2024). GIS based multicriteria decision making to identify regional groundwater potential zones: A critical review. *Sustainable Water Resources Management, 10*(61), 1–12. https://doi.org/10.1007/S40899-024-01080-9/METRICS

Kodihal, S., Akhtar, M. P., & Maurya, S. P. (2024). Hydroclimatic projection: Statistical learning and downscaling model for rainfall and runoff forecasting. *Journal of Water and Climate Change, 15*(2), 759–772. https://doi.org/ 10.1007/s40899-024-01080-9

Kumar, A., Shirin, S., Ansari, M. I., Pandey, G., Sharma, S. N., & Yadav, A. K. (2022). Modeling and simulation of environmental systems: A computation approach. In Maurya, S. P., Yadav, A. K. and Singh, R. (eds), *Modeling and Simulation of Environmental Systems,* 153–176, https://doi.org/10.1201/9781003203445-10 CRC Press.

Lentswe, G. B., & Molwalefhe, L. (2020). Delineation of potential groundwater recharge zones using analytic hierarchy process-guided GIS in the semiarid Motloutse watershed, eastern Botswana. *Journal of Hydrology: Regional Studies, 28,* 100674. https://doi.org/10.1016/J.EJRH.2020.100674

Makropoulos, C. K., & Butler, D. (2006). Spatial ordered weighted averaging: Incorporating spatially variable attitude toward risk in spatial multicriteria decision-making. *Environmental Modeling & Software, 21*(1), 69–84. https://doi.org/10.1016/j.envsoft.2004.10.010

Malczewski, J. (1999). *GIS and multicriteria decision analysis.* 392. www.wiley.com/en-us/GIS+and+Multicriteria+Decision+Analysis-p-9780471329442

Malczewski, J., Chapman, T., Flegel, C., Walters, D., Shrubsole, D., & Healy, M. A. (2003). GIS–Multicriteria evaluation with ordered weighted averaging (OWA): Case study of developing watershed management strategies, *35*(10), 1769–1784. https://doi.org/10.1068/A35156

Malczewski, J., & Liu, X. (2014). Local ordered weighted averaging in GIS-based multicriteria analysis. *Annals of GIS, 20*(2), 117–129. https://doi.org/10.1080/19475683.2014.904439

Memarian, H., Balasundram, S. K., Talib, J. B, Sung, C. T. B., Sood, A. M., & Abbaspour, K. (2012). Validation of CA-markov for simulation of land use and cover change in the Langat Basin, Malaysia. *Journal of Geographic Information System, 2012*(06), 542–554. https://doi.org/10.4236/JGIS.2012.46059

Mishra, S.K. & Singh, V.P. (2003). Soil Conservation Service Curve Number (SCS-CN) Methodology, *42*, Water Science and Technology Library, Springer, Netherlands, Dordrecht. https://doi.org/10.1007/978-94-017-0147-1

Mokarram, M., Negahban, S., Abdolali, A., & Ghasemi, M. M. (2021). Using GIS-based order weight average (OWA) methods to predict suitable locations for the artificial recharge of groundwater. *Environmental Earth Sciences, 80*(12), 428. https://doi.org/10.1007/s12665-021-09719-y

Mussa, C. C., Gomaa, D., Rowley, D. D., Schmidt, U., Ginier, E., & Strickland, S. L. (2020). AARC Clinical Practice Guideline: Management of Adult Patients with Tracheostomy in the Acute Care Setting. *Respiratory Care*, *66*(1), 156–169. https://doi.org/10.4187/respcare.08206

Patil, D., & Gupta, R. (2023). GIS-based multicriteria decision-making for ranking potential sites for centralized rainwater harvesting. *Asian Journal of Civil Engineering*, *24*(2), 497–506. https://doi.org/10.1007/s42107-022-00514-z

Patil, D., Kumar, G., Kumar, A., & Gupta, R. (2023). A systematic basin-wide approach for locating and assessing volumetric potential of rainwater harvesting sites in the urban area. *Environmental Science and Pollution Research*, *30*(6), 14707–14721. https://doi.org/10.1007/S11356-022-23039-Z/TABLES/2

Rahman, M. M., & Szabó, G. (2022). Sustainable urban land-use optimization using GIS-based multicriteria decision-making (GIS-MCDM) approach. *ISPRS International Journal of Geo-Information*, *11*(5), 313. https://doi.org/10.3390/ijgi11050313

Rastogi, U., Kumar, S., Maurya, S. P., & Yadav, A. K. (2025). Air pollution forecasting using genetic algorithm and LSTM deep learning technique. *Aerosol Science and Engineering*, 1–10. doi: 10.1007/s41810-025-00343-y.

Saad, M., Nofal, E., Abdelmonem, Y., & Riad, P. (2024). Application of decision support system/remote sensing/GIS techniques in groundwater recharge assessment. *Water Practice and Technology*, *19*(9), 3721–3743. https://doi.org/10.2166/WPT.2024.193

Saaty, T. L. (1977). A scaling method for priorities in hierarchical structures. *Journal of Mathematical Psychology*, *15*(3), 234–281. https://doi.org/10.1016/0022-2496(77)90033-5

Saaty, T. L. (1989). Decision making, scaling, and number crunching. *Decision Sciences*, *20*(2), 404–409. https://doi.org/10.1111/j.1540-5915.1989.tb01887.x

Seland, Ø., Bentsen, M., Olivié, D., Toniazzo, T., Gjermundsen, A., Graff, L. S., Debernard, J. B., Gupta, A. K., He, Y. C., Kirkevåg, A., Schwinger, J., Tjiputra, J., Schanke Aas, K., Bethke, I., Fan, Y., Griesfeller, J., Grini, A., Guo, C., Ilicak, M., … & Schulz, M. (2020). Overview of the Norwegian earth system model (NorESM2) and key climate response of CMIP6 DECK, historical, and scenario simulations. *Geoscientific Model Development*, *13*(12), 6165–6200. https://doi.org/10.5194/GMD-13-6165-2020

Sunkur, R., Kantamaneni, K., Bokhoree, C., Rathnayake, U., & Fernando, M. (2024). Mangrove mapping and monitoring using remote sensing techniques toward climate change resilience. *Scientific Reports*, *14*(1), 1–14. https://doi.org/10.1038/s41598-024-57563-4

Wu, W. T., Zhou, Y. X., & Tian, B. (2017). Coastal wetlands facing climate change and anthropogenic activities: A remote sensing analysis and modeling application. *Ocean & Coastal Management*, *138*, 1–10. https://doi.org/10.1016/J.OCECOAMAN.2017.01.005

Yadav, P., Ladha, S., Deshpande, S., & Curry, E. (2019). Computational model for urban growth using socioeconomic latent parameters. *Lecture Notes in Computer Science (Including Subseries Lecture Notes in Artificial Intelligence and Lecture Notes in Bioinformatics)*, *11329 LNAI*, 65–78. https://doi.org/10.1007/978-3-030-13453-2_6

Yager, R. R. (1988). On ordered weighted averaging aggregation operators in multicriteria decisionmaking. *IEEE Transactions on Systems, Man and Cybernetics, 18*(1), 183–190. https://doi.org/10.1109/21.87068

Youssef, Y. M., Gemail, K. S., Atia, H. M., & Mahdy, M. (2024). Insight into land cover dynamics and water challenges under anthropogenic and climatic changes in the eastern Nile Delta: Inference from remote sensing and GIS data. *Science of The Total Environment, 913*, 169690. https://doi.org/10.1016/J.SCITOTENV.2023.169690

Zhang, X., Song, Y., Nam, W. H., Huang, T., Gu, X., Zeng, J., Huang, S., Chen, N., Yan, Z., & Niyogi, D. (2024). Data fusion of satellite imagery and downscaling for generating highly fine-scale precipitation. *Journal of Hydrology, 631*, 130665. https://doi.org/10.1016/J.JHYDROL.2024.130665

5

Adaptive Modelling for Climate Change Action Plans: Principle and Practices

Ramesh Singh, Satya Prakash Maurya, Akhilesh Kumar Yadav, and Vijay P. Singh

5.1 Introduction

Owing to the lack of synchronized components and fine tuning, all studies on sustainable development based on scientific and natural phenomena appear to be less effective and call for a more responsive system. Because the industrial and environmental aims were not aligned, the prior methods failed to address urgent research needs. Nonetheless, it has often been asserted that upcoming projects will be sustainable and environmentally friendly.

The necessity to lessen humans' primary negative effects on nature to protect the natural environmental system, on the other hand, gave rise to social ecology. Likewise, the idiosyncrasies of social ecology's future predictions are governed by the specific circumstances of its birth. Any field of study aims to create new opportunities for mastering social and natural reality through its predictive functions. Determining the limiting values of society's future development in the use of any natural resource is the main goal of social ecology.

While the natural process of the self-regulating biosphere is used to fully ensure the integrity and fitness of the physical environment for life, its ability to maintain its integrity as a whole has been significantly diminished and, in some cases, has already been depleted. Thus, we may now discuss the necessity of understanding and intentionally using a brand-new set of laws that are fundamentally socio-natural and whose requirements must be met to maintain the integrity of the "society-nature" system.

5.1.1 Environmental Modeling Method

Over the years, significant progress has been made in understanding ecological and environmental issues, and the advancement of computing technology has allowed us to manage increasingly complicated mathematical

DOI: 10.1201/9781042013234-6

systems. Environmental technology, cleaner technology, ecological engineering, or eco-technology, in conjunction with environmental legislation, implementation constraints, policy and practices, regulation and governance, and other factors, are all significant components of the sustainability of ecosystem management with human interaction (Figure 5.1). Only the items and processes relevant to the issue under consideration should be the focus of ecological models. An ecological model never includes every detail for the area under consideration because doing so would detract from the map user's primary goal. Similarly, an excessive amount of unrelated data is removed from the model's primary goals (Reis & Nitter, 2013).

Environmental problems are extremely complex because of their high degree of uncertainty, variability in both temporal and spatial scales, potential irreversibility of damage, cross-sectoral nature, and multi-level spanning. Furthermore, solutions to environmental issues are composed of numerous interrelated ecological elements and are implemented inside and dependent upon social systems (Kirchke et al., 2017; Singh et al., 2022). Thus, it is necessary to handle the following inputs and their relationships.

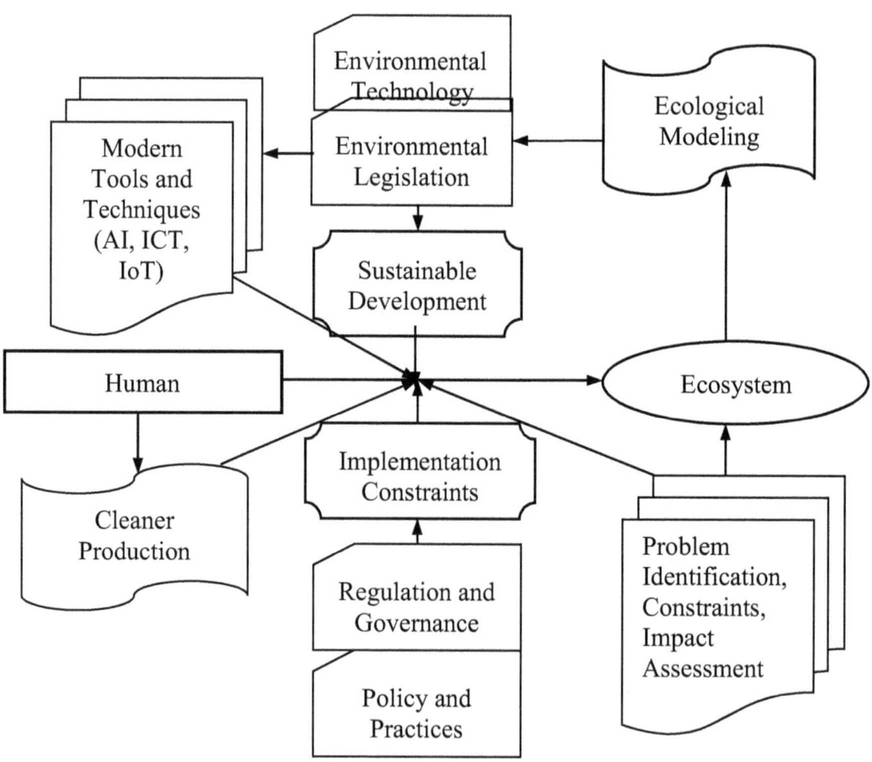

FIGURE 5.1
Using modeling and contemporary tools and strategies to control the environment.

5.1.2 Need for Adaptive Planning Due to Climate Change

With its effects on ecosystems, human health, water supplies, agriculture, infrastructure, and economic systems worldwide, climate change is one of the 21st century's most complicated and dynamic concerns. Climate change is defined by its long-term trajectory, nonlinear repercussions, and substantial uncertainties, in contrast to the static environmental concerns of the past (UN-SDG 13). These uncertainties are caused by the natural complexity of Earth's climate systems, socioeconomic trends, and variations in greenhouse gas emissions.

Conventional methods of climate planning frequently depend on strict deadlines, set assumptions, and one-dimensional future scenarios. However, given rapidly changing climate realities, such static models and linear planning frameworks are becoming increasingly less effective. Decision-makers want frameworks that are naturally adaptable, responsive, and iterative as new data become available and situations change quickly, whether as a result of sudden temperature shifts, unanticipated weather extremes, or tipping points such as polar ice melt or forest die-back. By switching from deterministic forecasting to dynamic decision-making, adaptive planning meets this demand. Developing strategies that can adapt to new information, stakeholder feedback, and observed environmental changes embraces uncertainty rather than attempts to eradicate it. Instead of optimizing for a single anticipated outcome, the main concept is to develop well-designed and flexible policies that function well in a variety of future circumstances.

This planning paradigm is especially important for areas and industries that are particularly susceptible to climate threats, such as coastal towns, rain-fed agricultural zones, and dry or mountainous regions. Adaptive planning in these situations allows governments and communities to:

- Interventions should be modified in light of ongoing environmental monitoring.
- Utilize predictive modeling to identify new hazards.
- Take stakeholder priorities and local knowledge into consideration.
- Re-evaluate tactics in light of emerging climate research or changes in society and politics.

Additionally, adaptive planning is in line with global frameworks that prioritize flexibility, inclusivity, and long-term resilience, such as the Sustainable Development Goals (SDGs), Paris Agreement, and Sendai Framework for Disaster Risk Reduction.

In this chapter, we examine how adaptive modeling—the application of scenario-based, modular, and responsive computational models—can be a fundamental component of climate action planning. Climate adaptation is a constant and well-informed process thanks to adaptive models that integrate research, systems thinking, and stakeholder input to guide decision-making under uncertainty and promote proactive learning.

5.1.3 Adaptive Modeling: What Is It?

A family of modeling techniques known as "adaptive modeling" is made to change when data, situations, and goals do. It is especially well-suited for complex systems, such as climate change. Adaptive models are iterative, responsive, and able to include real-time feedback, uncertainty, and changing system behaviors, in contrast to traditional static models that are frequently calibrated to a preset dataset or a predefined scenario. Fundamentally, adaptive modeling is a planning philosophy and methodological framework. Enabling users to evaluate the resilience of interventions, simulating a variety of potential futures, and modifying methods in response to observed outcomes or new information aids in decision-making in the face of ambiguity.

5.1.4 Design Iteration

The adaptive models are designed to be improved over time. The model can be updated and recalibrated as new data become available, whether via remote sensing (RS), climate sensors, stakeholder inputs, or advances in science. A system based on learning was produced through this continuous cycle of observation, model modification, and decision-making.

5.1.4.1 Flexibility in Scenarios

Several scenarios are frequently included in adaptive modeling frameworks to depict various potential future states. For instance, models may replicate policy responses, urban growth patterns, or various paths of greenhouse gas emissions (e.g., SSPs/RCPs). The goal of the model is to stress-test tactics under a variety of possible trajectories, rather than to forecast a single future.

5.1.4.2 Loops of Feedback and Learning

Adaptive models can dynamically alter system behavior over time by integrating feedback mechanisms. For instance, a drought management model would incorporate feedback from reservoir levels, water usage trends, and the implementation of governmental policies. This enables the system to react to water scarcity in real time in an adaptive manner.

5.1.4.3 Integration of Multidisciplinary

Adaptive models incorporate socioeconomic, ecological, and physical factors. For example, a coastal flooding model might provide a comprehensive understanding of vulnerability and response tactics by connecting sea-level rise estimates with land use planning, population movement, and infrastructure resilience.

5.1.4.4 Participation of Stakeholders

Input from scientists, politicians, local communities, and other stakeholders is incorporated into effective adaptive modeling. By incorporating models into institutional and societal decision-making processes, this participatory feature not only increases the legitimacy and applicability of model outputs but also strengthens adaptive capability.

A growing number of sophisticated computational tools are supporting adaptive modeling, such as ABMs or agent-based models, which are used to model both individual and group behavior. Models of system dynamics (SD) to comprehend intricate feedback systems, machine learning (ML) models for identifying predictive patterns, spatially explicit analysis, and geospatial integrated models. Adaptive modeling offers an organized approach to investigating "what if" scenarios, determining low-regret tactics, and updating policies on a regular basis if conditions or new threats arise in the framework of climate change action plans. By transforming planning from a one-time event to an ongoing process, this method promotes governance and system resilience.

5.2 Chapter Structure and Goals

This chapter establishes a context for comprehending how adaptive modeling functions as a tactical instrument in the creation and development of climate change action plans. Given that the effects of climate change are multifaceted, unpredictable, and ever-changing, this chapter attempts to provide readers with a morally sound and useful framework for integrating adaptive models into planning and policy. The chapter's main goals are:

- Recognize the shortcomings of conventional, static planning techniques in mitigating climate threats.
- Understand the fundamental ideas and goals of adaptive modeling.
- Determine the tools and frameworks for adaptive modeling employed in climate-related decision-making.
- Recognize the importance of iterative learning and stakeholder engagement in adaptive planning.
- Examine case studies that demonstrate effective adaptive modeling implementations in various settings.

5.3 Foundations of Theory

Adaptive modeling is based on a broad range of interdisciplinary theories and concepts that help us understand how complex systems operate, how to handle uncertainty, and how to make decisions in the face of change. The main theoretical tenets supporting adaptive modeling in the context of climate change are introduced in this section.

5.3.1 Adaptive Modeling's Main Challenges and Possibilities

5.3.1.1 The Socio-Environmental Model's Dynamic

Even integrating concepts, information, and techniques from other fields within environmental research necessitates the construction of bridges, but tying social and natural science together involves integrating disparate worlds. While dynamics are a fundamental issue for environmental modelers, the basic philosophy concentrates on comprehending the sensations or relationships of a specific moment or period.

5.3.1.2 Ambiguity about the Definition of Data

The validity, quality, and significance of the data, as well as the methods for gathering and analyzing data, might all be unclear due to differences in epistemological backgrounds.

5.3.1.3 Absence of Accepted Standards for Cooperation

Though there has not been much progress thus far, there has been a growing awareness of the absence of transparency standards and data and model code sharing in all scientific domains (Baker, 2016). Sharing well-documented data and model codes is still not a common practice across all sectors (Janssen, 2017; Stodden et al., 2018).

5.3.1.4 Uncertainty Modeling

Although data, model structure, and model parameters are frequently cited as the primary causes of uncertainty, there are other sources of uncertainty that affect every decision made during the modeling process as well as every intermediate and ultimate result produced. Mostly quantitative in nature, uncertainty assessment has been limited to a more focused examination of the uncertainty surrounding model parameters, inputs, and other data. The definition of purpose and objectives, problem formulation, and model conceptualization to explore the proper challenge, parameters and boundary conditions, verification and validation, including its qualitative aspect, model coding, and modeling workflows employed are some of the modeling phases in which uncertainty arises. The importance of using a more comprehensive approach to evaluating uncertainty has recently gained greater emphasis. An overview of the possible sources of modeling uncertainty investigated using various techniques is given in Table 5.1.

5.3.2 Comprehending Environmental System Modeling's Uncertainty and Variability

Forecasting, management, and policy depend on environmental system models, including those related to hydrology, biogeochemistry, climate impacts, water resources, and ecology. However, two related but different ideas–uncertainty (lack of knowledge about the system, model, or future drivers) and variability (natural, frequently irreducible changes in a system)–shape their outputs. For reliable forecasts and creating flexible, risk-aware judgments, it is essential to comprehend and distinguish between these ideas. Both the taxonomy of uncertainty and useful techniques for measuring and expressing it have been highlighted in important reviews and syntheses. Moreover, nonlinear dynamics, delayed feedback, and interactions across spatial and temporal dimensions all contribute to the inherent complexity of climate systems. There are multiple causes of uncertainty in climate modeling: i). Natural fluctuations, such as cycles of the El Niño-Southern Oscillation (ENSO); ii. constraints on data, particularly in the Global South; and iii). assumptions for modeling, and iv). Unknown future human behaviors (e.g., legislative changes, land use, and energy use).

In general, there are two categories of uncertainty that are particularly significant: i). Epistemic doubt is brought about by a lack of information or expertise. ii). Aleatory uncertainty is caused by a system's intrinsic variability or randomness. However, uncertainty and unpredictability come from various causes.

5.3.3 Uncertainty and Variability Sources

Uncertainty is regularly categorized in the literature into several interrelated categories:

Stochastic fluctuations resulting from initial conditions or intrinsic system randomness are known as internal (natural) variability and are significant for extreme occurrences and short time frames. The uncertainty concerning parameter values within a selected model structure is known as the parameter uncertainty. The most difficult to measure is structural (model) uncertainty, which is the insufficiency or incompleteness of the model form (omitted processes, approximations). Input/forcing uncertainty refers to mistakes or unknowns in future driving scenarios and observational inputs (e.g., pollution and land use).

TABLE 5.1
Climate change uncertainty and adaptive modeling

Title	Reference	Objective	Approach	Uncertainty addressed	Key findings	Strengths	Limitations	Relevance for adaptive modeling	Gaps
Climate adaptation as a control problem.	Herman et al. (2020).	Classify dynamic planning approaches by uncertainty handling	Control theory viewpoint; dynamic planning taxonomy	Parametric, stochastic, partially observed uncertainty	Viewing adaptation as a control problem clarifies tradeoffs between proactive vs reactive options and how uncertainty propagates.	Theoretical framing linking control and adaptation.	Less direct hands-on guidance for large multimodel ensembles.	Provides a mathematical basis for adaptive update rules and control-informed monitoring thresholds.	Implementation guidance needed for high-dimensional climate models and socio-economic coupling.
Method uncertainty in precipitation projections.	Uhe et al. (2021).	Demonstrate importance of method/processing choices (bias correction etc.)	Sensitivity experiments across methods	Methodological/processing uncertainty; structural	Different methodological choices produce large differences in confidence statements for precipitation — methoduncertainty is essential to quantify.	Empirical demonstration; variable-specific.	Focused on precipitation; generalization required.	Adaptive models must include method/process uncertainty in propagation to impacts and triggers.	Standardized frameworks for representing method uncertainty across impact chains.
Decision-making under deep uncertainty — review.	Stanton & Stirling (2021).	Synthesize DMDU methods and applications to long-term climate decisions	Review of Robust Decision Making (RDM), Decision Scaling, Adaptive Pathways, scenario methods	Deep (epistemic) uncertainty; scenario uncertainty; parametric	DMDU approaches help identify robust policies across wide scenario sets, and model diversity must be explicitly represented.	Broad method comparison; high policy relevance.	Less focus has been placed on coupling with process models or explicit structural error quantification.	Useful toolbox for adaptive modeling — chooses method to match decision time horizon and monitoring ability.	Better standards are needed for combining model ensembles with DMDU workflows.

Partial identification / deep uncertainty framing.	Manski et al. (2021).	Statistical framing for limited identifiability in climate model outputs	Econometric partial identification; bounding predictions	Epistemic / method uncertainty; model identification limits	Proposes framing ensemble outputs as bounds instead of point probabilities where identification fails.	Rigorous treatment of inference limits; useful for decision makers.	Conceptual + technical; applied examples limited.	Encourages adaptive strategies that acknowledge wide credible bounds (useful for conservative triggers).	Practical methods to translate bounds into actionable triggers and monitoring strategies.
Techniques for DMDU.	NOAA DMDU report (2021).	Review DMDU techniques and potential NOAA use	Technical report summarizing methods & use cases	Deep uncertainty; scenario & model uncertainty	Catalogues DMDU methods (RDM, adaptive pathways, etc.) and their strengths for operational use.	Practitioner-oriented guidance.	US-agency focus; evolving field.	Good source for operationalizing adaptive monitoring and decision thresholds in agencies.	Implementation case studies across sectors and scales.
Quantifying uncertainty sources of future climate.	Wu et al. (2022).	Partition uncertainty into forcing, scenario, model, internal variability	Statistical decomposition across ensembles (CMIP)	Forcing (scenario), model structural, internal variability	Model structural uncertainty is a major contributor to projection spread (often second after scenario depending on timescale/ variable).	Quantitative partitioning; uses multi-model ensembles.	Dependent on available ensemble selection; attribution can be sensitive to method.	Indicates where adaptive models should target uncertainty reduction vs robust decision design.	Need methods to separate structural vs shared biases and how to propagate into impact models.
Novel ensemble framework for climate risk assessment.	Dawkins et al. (2023).	Integrate ensembles into CLIMADA platform for risk assessment	Framework + application examples	Ensemble & scenario uncertainty; compound risks	Presents practical ensemble integration for risk assessments, showing how ensemble spread affects risk metrics.	Practical, implementable approach for impact/risk users.	Platform-specific examples; generalization policy dependent.	Useful template for coupling climate ensembles to adaptation appraisal and stress tests.	Broader adoption across sectors; linking to DMDU workflows.

(continued)

TABLE 5.1 (Continued)

Climate change uncertainty and adaptive modeling

Title	Reference	Objective	Approach	Uncertainty addressed	Key findings	Strengths	Limitations	Relevance for adaptive modeling	Gaps
DMDU and IPCC engagement.	Lempert et al. (2024).	Discuss evolving use of DMDU in IPCC assessments and practice	Review & institutional analysis	Deep uncertainty; framing/ communication uncertainty	Argues for greater incorporation of DMDU into assessment processes; highlights institutional barriers.	Connects scientific methods to assessment practice.	Institutional focus; fewer technical recipes.	Important for translating model uncertainty into adaptive policy frameworks and for communicating model limits.	Research on pragmatic protocols for IPCC/country use of DMDU + ensembles.
Large ensembles as test beds.	Lehner et al. (2024).	Role of large initial-condition ensembles for applied climate risk.	Review & examples using large GCM ensembles	Internal variability; parametric; compound event uncertainty.	Large ensembles are valuable to sample variability and test adaptation strategies under many realizations.	Emphasizes test-bed utility and compound events.	Computational cost; limited structural diversity if single model family used.	Use large ensembles to design monitoring/ triggers in adaptive pathways and to stress-test decisions against variability.	Combine with multi-model large ensembles to address both internal and structural uncertainties.
Lessons from adaptive pathways studies.	Haasnoot et al. (2024).	Evaluate use/value/ weaknesses of adaptive pathways for long-term planning	Synthesis of pathway case studies; conceptual analysis	Deep uncertainty; scenario & structural uncertainty; decision timing	Adaptive pathways are useful to plan sequences of actions with trigger points but require explicit framing of uncertainties and monitoring plans.	Practical, policy-oriented lessons; recent synthesis.	Case-study bias; limited formal quantification.	Directly informs design of adaptive policies that can respond to model uncertainty — emphasizestriggers, monitoring and robustness.	Need more quantitative integration with model ensembles and probabilistic updates.

Methodological uncertainty: decisions that affect outputs through data processing, calibration, bias correction, and downscaling. By avoiding excessive dependence on point forecasts, adaptive modeling accepts these uncertainties. Rather, it employs probabilistic methods, scenario ensembles, and decision analysis strategies that hold up well for a wide range of likely futures.

5.3.4 Techniques for Measuring and Diagnosing Uncertainty

A sophisticated toolset is currently available to identify and measure various sources of uncertainty. The most popular methods are as follows:

Global sensitivity analysis (SA) (Sobol, variance-based indices, Morris screening) finds important parameters and inputs and directs data collection and model simplification. Convergence checks, appropriate experimental design, and interpretation based on environmental model goals have all been emphasized in systematic reviews.

Ensembles and multi-model inter-comparisons: Single-model large ensembles and multi-model ensembles (MIPs) each sample internal variability and structural diversity independently; when combined, they aid in dividing uncertainty across sources and stress-testing adaptation alternatives. To cover both structural spread and internal sampling, recent work emphasizes the necessity of combining the two ensemble types.

Equifinality/GLUE and Bayesian methods: While formal Bayesian (MCMC, BMA) methods quantify posterior distributions and can try to separate forcing, parameter, and structural errors, informal Monte-Carlo approaches such as GLUE embrace equifinality by accepting a variety of behavioral parameter sets. Comparative research demonstrates that, whereas formal and informal methods frequently yield comparable uncertainty ranges, their interpretations and presumptions vary.

Breakdown and attribution of uncertainty: Statistical partitioning techniques measure relative contributions (e.g., variance breakdown between models, scenarios, and internal variability), which is useful for setting priorities for adaptation and research alternatives.

Iterative uncertainty quantification: Rather than waiting until the conclusion of model creation, emerging processes recommend assessing uncertainty early in the model development process to influence data gathering, model choice, and decision relevance.

5.3.5 Spreading into Decision and Effect Models

Impact chains, such as climate → hydrology → flood risk → economic loss, must be used to spread uncertainty. The literature cautions against understating method uncertainty, which can dominate the ultimate uncertainty in impacts and includes bias correction, downscaling, and impact model selections. Adaptive pathways, decision making under deep uncertainty (DMDU), and robust decision making (RDM) are commonly suggested frameworks for connecting quantifiable uncertainties to action sequences pertinent to policy. Nevertheless, a number of obstacles impede advancement:

Because numerous models make similar assumptions and because measurements might not be sufficient to distinguish between different structures, structural model errors are still difficult to detect and minimize.

Representativeness and computational cost: Full sampling (global SA, MCMC, and multi-model big ensembles) is costly, and the results may be skewed by ensemble design compromises.

Methodological transparency: Uncertainty statements are significantly affected by decisions made about preprocessing, calibration, and performance indicators; standardized reporting is currently lacking.

Communication and application: It can be difficult to translate complicated, occasionally non-probabilistic, uncertainty representations (bounds, plausible ranges) into decision rules, particularly for stakeholders who are not familiar with probabilistic nuance.

Additionally, the following factors may be considered in the study and practices:

Operational indicators: To transition AST from metaphor to operational diagnostics and create reliable, quantifiable indicators for resilience capacity, adaptive-cycle phases, and cross-scale vulnerability.

Integrated socio-ecological monitoring aims to balance early warning detection, learning experiments, and decision-trigger thresholds by specifically designing monitoring systems to support adaptive governance.

Institutional innovation: Research and test flexible, polycentric governance structures that allow for controlled experimentation while adhering to accountability and legal requirements.

Connecting quantitative models and governance: Test adaptive policies in the face of uncertainty and find solid tactics across conceivable futures by combining stakeholder-driven scenarios with mechanistic SES models.

5.3.6 Environmental Research with Adaptive Systems Approaches

Conceptual frameworks and comparative case studies: A large portion of the literature on adaptive systems maps adaptive cycles, thresholds, and governance arrangements using conceptual syntheses and comparative case studies (e.g., studies of fisheries, wetlands, and fire regimes). This offers compelling causal explanations and testable hypotheses.

Adaptive dynamics are formalized, and test scenarios for reorganization, tipping, or resilience-building interventions are tested using coupled models (agent-based, system-dynamics, and network models) in quantitative systems modeling.

Socio-institutional analysis: To determine the obstacles and facilitators of adaptive governance (participation, information flows, and legal limits), governance-focused research employs institutional analysis, co-production studies, and policy evaluation.

Experimental and monitoring designs: Monitoring and decision-trigger rules are increasingly incorporated into adaptive-management programs, and enhanced statistical techniques are employed to identify indicators of state change and learning efficacy.

These methods can be used in various applications.

Ecosystem management: Fisheries, watersheds, and wetland restoration make extensive use of adaptive system-informed techniques (adaptive management and resilience assessment). When necessary, they assist in locating managerial levers that support safe transformation or uphold desired regimes.

The design of nature-based solutions and adaptation techniques that expressly prioritize redundancy, variety, and institutional flexibility to face uncertain futures is guided by the notions of resilience and AST.

Cross-scale governance: Effective adaptations frequently rely on layered governance arrangements that connect local learning to larger-scale resources and supportive policies in addition to local activities. However, cross-scale coordination failure is a frequent obstacle.

Rather than being optimal under specific assumptions, adaptive models frequently seek to support resilient and adaptable policies. These goals are frequently supported by tools such as dynamic adaptive policy pathways (DAPP), info-gap theory, and robust decision-making (RDM).

- Resilience is the ability of a system to withstand shocks and to maintain its fundamental structure and function.
- The ability to adjust one's route or plan in response to changing circumstances is referred to as flexibility.
- Maintaining performance in a variety of circumstances without having to optimize for any one future is meant by robustness.

Adaptive planning is both social and technological. It calls for collaborative issue framing among scientists, politicians, communities, and other stakeholders, as well as inclusive discourse and shared learning. Involving stakeholders guarantees that models are based on local reality, represent community values, and have a higher chance of producing workable endorsed solutions. Co-design workshops, serious games, and group model creation are examples of participatory modeling techniques that are essential for including a variety of viewpoints in the modeling process.

5.4 Fundamentals of Flexible Modeling

Adaptive modeling offers a useful framework for handling the complexity and uncertainty in climate change planning. It places more emphasis on adaptability, learning, and resilience than on predetermined forecasts. The fundamental ideas governing the creation and use of adaptive models for climate action are described in this section.

5.4.1 Development of Incremental and Iterative Models

The process of adaptive modeling is ongoing rather than a one-time event. Over time, the model changes as a result of: i). Phased development begins with basic prototypes and adds details over time (ii). ongoing calibration, which involves improving the model whenever fresh information or input becomes available; and iii). Learning cycles: using results to update presumptions or frameworks. This idea encourages adaptive management, in which plans and policies are evaluated and improved in the real world as opposed to waiting for "perfect" forecasts.

5.4.2 What-If Research and Scenario Planning

Adaptive models use scenario analysis to examine several likely futures rather than relying only on one prediction. Among them are the following: i). Trajectories of high and low emissions (e.g., SSPs/RCPs) ii. distinct energy or land use paths; iii). varying degrees of technological or institutional capability. By testing plans under various scenarios, what-if analysis enables stakeholders to find solutions that are: i). Robust: able to function effectively in a variety of situations; ii). Flexible: able to be changed in the middle; iii). Low-regret: Provides advantages even if there are no dangers.

5.4.3 Learning Systems and Feedback Loops

Feedback is a common feature of dynamic real-world systems and includes both balancing (negative) and reinforcing (positive) loops. Such feedback is included in adaptive models to mimic real-time responses, such as i). behavioral shifts brought about by climate policy; ii). tipping points or ecosystem thresholds; and iii). degradation of resource systems or infrastructure over time. The model's realism and policy relevance were improved by including feedback. Additionally, it enables models to gain knowledge from observations and assessments, enhancing decision-making in every cycle.

5.4.4 Combining Data from Multiple Scales and Domains

Processes in many temporal and spatial dimensions both influence and are influenced by climate change. Integrated adaptive models: i). local to global data (e.g., projections of global temperatures versus home susceptibility); ii. information from several sectors (e.g., agriculture, health, energy, and water); iii. inputs that are both quantitative and qualitative (such as surveys, stakeholder narratives, and data from the RS). A systems view of climate risk, and adaptation is made possible by this integrative approach, which considers trade-offs and interdependencies.

5.5 Tools and Techniques for Adaptive Modeling

Numerous methods of science, computer modeling, artificial intelligence (AI), and geographic analysis have been used in adaptive modeling. Depending on the kind of decision-making help required, the system under study, and data availability, each technique has specific advantages.

5.5.1 System Dynamics Models

A technique for comprehending intricate, feedback-driven systems using stocks, flows, feedback loops, and time delays is called System Dynamics Modeling (SDM). SDM was developed by Jay Forrester in 1961 and is now a commonly used approach in climate change research to examine the relationships between social, economic, and environmental subsystems. It enables scholars and decision-makers to study long-term dynamic behavior, model policy situations, and comprehend how feedback impacts interactions between the climate and society. SDMs are used in a variety of areas in climate change research, including energy transition, population-resource feedback, adaptation planning, emission reduction policies, carbon cycle dynamics, and socio-ecological system resilience.

5.5.2 Agents-Based Models

In the context of climate change modeling, Agent-Based Models (ABMs) are computer programs that represent individual agents, such as households, farms, businesses, or governments, who interact and make decisions in their own socioeconomic and environmental environments. ABMs, as opposed to aggregate or top-down models, capture heterogeneity, adaptation, and emergent behavior, enabling researchers to investigate how complex system-wide climate outcomes can be produced by micro-level actions. In fields where feedback between human behavior and environmental reactions is nonlinear and context-specific, such as land-use change, energy transitions, adaption behavior, and policy diffusion, they are especially useful. ABMs may show how policies, risk perceptions, and collaboration change under climate stress by combining behavioral rules, social influence, and learning mechanisms. ABMs are essential for developing adaptive and socially conscious climate mitigation and adaptation methods because they offer strong insights into the dynamics of human–environment interactions, despite difficulties with calibration, validation, and data intensity. Tools: Repast, GAMA, and NetLogo.

5.5.3 Integrated Assessment Models

To examine how human and natural systems interact in the context of climate change, Integrated Assessment Models (IAMs) are extensive frameworks that integrate knowledge from several fields, including economics, energy systems, land use, and climate research. To examine long-term climate–economy pathways and the effects of mitigation and adaptation efforts, IAMs link greenhouse gas emissions, atmospheric processes, temperature responses, economic costs, and policy alternatives. While more recent models such as REMIND, IMAGE, and GCAM provide in-depth depictions of energy technologies, land-use dynamics, and regional heterogeneity, classic models such as DICE, FUND, and PAGE concentrate on global cost-benefit optimization. IAMs serve as the foundation for the IPCC's major assessments, which guide targets such as keeping warming to 1.5 or 2 degrees Celsius. Despite their policy significance, IAMs are criticized for underrepresenting social and ecological aspects, simplifying complicated feedback, and presuming rational decision-making. However, they continue to be essential instruments for scenario analysis, assessing carbon pricing and net-zero approaches, and connecting scientific knowledge to climate-policy choices. Despite their frequent global scope, IAMs are increasingly being downscaled for national or regional planning.

5.5.4 Using Geospatial Technology to Model Climate

By offering spatially detailed data that are necessary for comprehending, tracking, and forecasting climate patterns, geospatial technology, which includes geographic information systems (GIS), RS, and global positioning systems, plays a crucial role in climate modeling and research. RS uses high-resolution satellite observations to record important variables that feed into and evaluate climate models, including vegetation cover, sea level rise, land surface temperature, and air composition. The integration, visualization, and spatial analysis of climate datasets are made possible by GIS, which makes it easier to map exposure, risk, and susceptibility at different scales. When combined, these technologies improve climate models' precision and geographic detail, facilitate downscaling of global climate projections to local and regional contexts, and allow for real-time monitoring of climate-related phenomena such as deforestation, floods, and droughts. Geospatial tools are essential for data-driven decision-making, climate impact assessment, and adaptive resource management in a changing climate because they bridge the gap between large-scale climate simulations and reality on the ground.

5.5.5 Predictive Adaptation using Artificial Intelligence and Machine Learning

Predictive adaptation—the early detection of new threats and prompt adoption of countermeasures—is made possible by ML and AI, which profoundly change how societies expect and react to climatic and environmental changes. To identify early warning signs of extreme events, predict vulnerability hotspots, and maximize adaptive responses in industries such as agriculture, water management, and disaster risk reduction, these technologies make use of large, diverse datasets (such as climate projections, RS, and socioeconomic indicators). While AI-driven decision systems and reinforcement learning facilitate dynamic, real-time adaptation planning, deep learning models can capture complicated feedback and nonlinear interthereby actions, increasing the accuracy of flood and drouheat wave heatwave predictions. However, to ensure that AI-enhanced adaptation remains transparent, equitable, and in line with local knowledge and priorities, effective deployment necessitates addressing issues of interpretability, data quality, and ethical governance.

5.6 Constructing and Applying Adaptive Models

The process of creating an adaptive model for climate change planning is intricate and iterative and requires a delicate balance between stakeholder demands, scientific rigor, and decision-making deadlines. The main steps involved in creating and implementing these models are described in this section.

5.6.1 Framing the Issue and Establishing Goals

Clear problem-framing and objective setting are the first steps in effective adaptive modeling. These establish the boundaries of the system, the main forces behind change, the priorities of stakeholders, and the choices the model intends to support. Recognizing that uncertainties—scientific, socioeconomic, and institutional—cannot always be avoided but must instead be publicly portrayed and handled is necessary for problem framing in the context of complex and unpredictable environmental systems. Thus, adaptive models are designed to promote learning, flexibility, and decision-making in the face of uncertainty, rather than just forecasting results. As new data, information, and stakeholder viewpoints become available, the objectives themselves must be flexible and iterative. To ensure that the model represents real-world issues and practical policy requirements, this method necessitates the co-production of knowledge among scientists, decision-makers, and end users. Adaptive models are living instruments that are continuously improved to increase resilience, direct sound decisions, and effectively respond to shifting environmental and societal situations. This is made possible by well-framed challenges and precisely stated flexible objectives.

Every attempt at adaptive modeling starts with a precise specification of:

- The issue that needs to be resolved (such as crop failure, coastal erosion, or urban heat risk),
- The parties concerned, and the important choices that need to be backed.
- A well-framed model is scalable, sensitive to real-world constraints, and policy-relevant.

5.6.2 Information Gathering and Combination

Continuous, high-quality data collection and smooth integration across various sources are essential for effective adaptive modeling in order to guarantee that models continue to be sensitive to changing socioeconomic and environmental conditions. To capture both system states and driving variables, data are collected from a variety of sources, including citizen science, in situ monitoring networks, RS, and socioeconomic databases. Through standardized formats, metadata protocols, and real-time data assimilation algorithms that update model parameters and projections as new information becomes available, integration entails harmonizing diverse datasets. Adaptive models can recognize changes, adjust predictions, and assist in making decisions in the face of uncertainty owing to this dynamic data–model interaction. Nonetheless, issues of data governance, temporal and spatial resolution discrepancies, and data interoperability still exist.

Therefore, maintaining adaptive models that can learn, adapt, and successfully inform resilience-based environmental management requires strengthening the data infrastructure, interoperability standards, and participatory monitoring systems.

Adaptive models use a variety of datasets:

- climate information (such as forecasts for temperature and precipitation),
- socioeconomic metrics (such as livelihoods and demography),
- environmental information (such as water availability and land cover),
- institutional information (such as regulatory criteria and policy deadlines).

For a model to be resilient, data must be integrated across scales (local to global) and types (quantitative, qualitative, and remote sensing).

5.6.3 Structure and Design of the Model

It is necessary to incorporate flexibility, feedback, and learning mechanisms into the design and structure of adaptive models so that they can adapt to new information, system modifications, and policy interventions. Adaptive models are based on iterative updating, which uses feedback loops, Bayesian learning, or real-time monitoring to adjust predictions and decision rules as circumstances change. This is in contrast to static models, which assume fixed parameters and relationships. Typically, their architecture comprises modular elements (such as policy, socioeconomic, and environmental subsystems) that are connected by dynamic interactions and pathways for the spread of uncertainty. Transparency, scalability, and responsiveness are prioritized to ensure that model outputs are still applicable in the face of extreme uncertainty. To ensure that the model not only replicates SD but also facilitates ongoing learning, policy modification, and resilience in intricate environmental and socio-ecological contexts, an effective adaptive model design also incorporates mechanisms for stakeholder involvement and scenario testing.

Important design choices include the following:

- modeling methodology type (e.g., ABM, SDM, ML),
- spatial and temporal resolution,
- important system elements and interactions,
- Simplified assumptions.

It is easier to update or replace individual components without affecting the entire system when components (such as climate, hydrology, and land use) are loosely connected in a modular architecture.

5.6.4 Verification and Adjustment

To ensure that adaptive models faithfully capture the behavior of real-world systems and can consistently guide decision-making in dynamic environments, calibration and validation are crucial procedures. To improve the model's fidelity while preserving parsimony, calibration involves methodically modifying the model's parameters so that the simulated outputs match the observed data. To account for parameter uncertainty and allow the model to learn from fresh data over time, it frequently employs ensemble techniques, Bayesian updating, or optimization. In contrast, validation assesses a model's credibility and generalizability by comparing its predicted performance and resilience with those of independent datasets or scenarios. Calibration and validation are repetitive processes rather than one-time events in adaptive modeling, in which models are updated continuously as new data and system feedback become available. This continuous procedure facilitates learning and model development, guaranteeing that the model will continue to be applicable as boundary conditions, SD, or management goals change. The foundation for reliable, adaptable, and evidence-based adaptive models that can direct resilient environmental and policy decisions in the face of uncertainty is provided through calibration and validation.

This stage guarantees realistic behavior in the model:

For model outputs to correspond with observable historical data, calibration entails modifying parameters. The model's capacity to replicate patterns in unknown or future circumstances was tested through validation.

5.6.5 Developing and Simulating Scenarios

Developing and simulating scenarios is essential for creating and implementing adaptive models, particularly in intricate environmental and climatic systems with high levels of uncertainty. Modelers may evaluate system behavior under a variety of assumptions owing to scenarios that offer quantitative paths and organized, believable narratives that encompass a range of future variables, including socioeconomic developments, policy responses, and climate trajectories. These situations were converted into dynamic outcomes through simulation, exposing thresholds, feedback, and possible tipping points. Instead of optimizing for a single anticipated outcome, this technique makes it possible to identify robust strategies that perform well across a wide range of possibilities. Adaptive models use these insights to integrate learning, flexibility, and iterative updating in response to new data. Scenario-based simulations are crucial tools for adaptive management, climate risk planning, and long-term sustainable policy formulation, because they direct the creation of decision triggers, monitoring indicators, and backup plans.

After validation, the model was used in various situations:

- climate projections (such as RCP 4.5 and RCP 8.5),
- policy changes (such as zoning regulations and the carbon price),
- socioeconomic patterns (such as migration and urbanization).

The aim is to evaluate the threshold effects, system behavior, and effectiveness of different approaches under different scenarios.

5.6.6 Involving Stakeholders and using Participatory Modeling

Building and implementing adaptable models, especially in contexts connected to the environment and climate, requires stakeholder engagement and participatory modeling. Participatory techniques include a variety of knowledge systems, priorities, and values in model design, scenario building, and policy testing by actively integrating stakeholders—including scientists, politicians, resource managers, and local communities—across the modeling process. This cooperative approach guarantees that adaptive strategies represent real-world restrictions and decision-making contexts, thereby improving model relevance, legitimacy, and usability. Additionally, iterative learning is promoted by participatory modeling, as stakeholders can examine the effects of different courses of action, comprehend feedback and uncertainties, and jointly design strong and adaptable solutions in the face of shifting circumstances. These methods enhance the design of monitoring, triggers, and adaptive management pathways by combining technical modeling with social learning. This helps close the gap between intricate model outputs and decisions that can be implemented with stakeholder assistance. Stakeholder input improves usefulness and trustworthiness. Among the methods are seminars for scenarios, simulated exercises or serious games, interactive dashboards, and building models in groups. Additionally, technical findings can be translated into practical tactics suited to institutional capabilities and community needs with the aid of participatory approaches.

5.6.7 Integration of Policy and Decision-Support

To convert adaptive models into workable plans for controlling environmental and climate hazards, decision-support and policy integration are essential. The analytical basis for investigating alternate possibilities, weighing trade-offs, and assessing the resilience of actions in the face of changing circumstances is provided by adaptive models that specifically consider uncertainty, feedback, and SD. Stakeholder goals, institutional capabilities, and governance frameworks must be closely aligned to integrate these models into decision-making processes. This is frequently accomplished through dynamic adaptive policy pathways, iterative learning, and participatory modeling. By ensuring that the model outputs are integrated into technical planning, policy design, monitoring, and adaptive management, it is possible to make timely adjustments as new information becomes available. Decision-makers can foresee hazards, find adaptable solutions, and promote resilience in intricate socio-environmental systems by incorporating adaptive modeling into policymaking processes.

The outputs of the model are converted into:

Criteria for making decisions (such as resilience ratings and cost-benefit ratios), spatial planning dashboards and maps, paths, or narratives that demonstrate techniques that alter over time. At this point, modeling is related to strategies for a strategic climate, setting priorities for investments, readiness for disasters, and frameworks for monitoring (such as indicators of climate adaptation).

5.6.8 Adaptive Learning, Monitoring, and Updating

The foundation of developing and implementing successful adaptive models, especially when it comes to environmental and climate systems, is monitoring, updating, and adaptive learning. Adaptive models are designed to work in situations of extreme uncertainty when assumptions about the model, external influences, and system behavior can all change over time. To identify departures from anticipated trajectories or new trends, monitoring entails methodically gathering pertinent data on system states, environmental factors, and model performance. To maintain the model's accuracy and relevance over time, updating entails changing the model's parameters, scenarios, or structure in response to fresh observations, revelations, or environmental changes. The iterative process by which stakeholders, modelers, and decision-makers analyze observed data, assess model outputs, and modify management tactics in response is known as adaptive learning. These procedures work together to create a dynamic feedback loop in which models inform decisions, decisions result in outcomes, results are tracked, and insights are used to improve strategies and models. In complex socio-ecological systems, this method promotes adaptable, evidence-based decision-making, strengthens resilience, and increases the robustness of policy interventions.

Lastly, models that are adaptive are dynamic. They have to be:

- Tracked for performance in the actual world,
- Updated with fresh information or understanding,
- Periodically reassessed to guarantee ongoing applicability.

One method for routinely reviewing policies in light of shifting thresholds and circumstances is the Dynamic Adaptive Policy Pathways (DAPP).

5.7 Integrating Policies and Providing Decision Support

The best results from the adaptive models are obtained by integrating them into actual planning procedures and policy cycles. This section describes how the modeling results help guide policy, facilitate uncertain decision-making, and promote institutional learning in climate action.

5.7.1 Translation Challenge: Moving from Models to Policy

One of the primary challenges in decision support and policy integration is the conversion of environmental and climatic models into workable policies. Although models—from integrated assessment and SD frameworks to process-based simulations—provide quantitative insights into system behavior, feedback, and scenario outcomes, their outputs are frequently fraught with technical complexities, assumptions, and uncertainties. However, to make decisions under political, social, and economic limitations, policymakers need clear, solid, and context-relevant guidance. In addition to effectively expressing uncertainty and framing outcomes in measures relevant to decisions, closing this gap requires co-producing knowledge through adaptive governance structures, iterative stakeholder engagement, and participatory modeling. The design of iterative feedback loops between modelers and decision-makers is crucial for successful model-to-policy translation, according to recent literature. This ensures that scientific insights are both institutionally actionable and scientifically credible, supporting flexible evidence-based policies in the face of profound uncertainty.

Converting intricate model outputs into understandable, policy-relevant insights entails the following:

- Simplifying outcomes without making the message too simple,
- Adapting information to the target audience (for example, community leaders versus policymakers),
- Connecting results to important planning choices (e.g., disaster preparedness, zoning, budgeting).

5.7.2 Deeply Uncertainty Decision Support

The goal of decision support under deep uncertainty (DSDU) is to assist policymakers when there are several likely outcomes, the future is extremely uncertain, and the probabilities are unclear or disputed. With its emphasis on resilience, flexibility, and iterative learning, DSDU differs from traditional risk-based decision-making, which depends on well-stated probabilities. This allows decision-makers to find solutions that work effectively in a variety of uncertain scenarios. Techniques such as scenario-based planning, dynamic adaptive policy pathways (DAPP), and robust decision making (RDM) are frequently used to create adaptive triggers, enable monitoring, and facilitate contingency planning. By offering transparent frameworks that integrate stakeholder interests, scientific information, and institutional restrictions, the DSDU promotes cooperation across sectors and scales in policy integration and guarantees that adaptation plans are resilient and practical in the face of unpredictability. This method, which helps governments and organizations traverse complicated socio-ecological systems without depending on exact projections, has been widely used for catastrophe risk reduction, energy policies, water management, and climate change adaptation.

Even in situations where the future is extremely unpredictable, planners can make judgments with the aid of adaptive models. Important methods include the following:

- Robust Decision Making (RDM): Finds solutions that work effectively in a variety of situations.
- Adaptive Policy Pathways (APP): Charts decision-making sequences associated with particular signals or thresholds.
- MCDA, or multi-criteria decision analysis, aids in balancing trade-offs between social, economic, and environmental objectives.

5.7.3 Integrating Models into Cycles of Policy

A crucial tactic for policy integration and decision assistance is to integrate models into policy cycles, which enables scientific insights to guide every phase of policy creation, execution, and assessment. Decision-makers can investigate scenario outcomes, foresee unforeseen repercussions, and analyze trade-offs across sectors by incorporating models, from agent-based and socio-ecological simulations to SD and integrated assessment models, into iterative policy processes. By connecting monitoring and learning to decision triggers, model-informed techniques promote adaptive governance, increase transparency, and ease stakeholder participation. To ensure that insights are useful without relying too heavily on forecasts, it is crucial to carefully consider model credibility, uncertainty communication, and institutional context while embedding models. By strengthening cross-sectoral policy integration, this strategy makes it possible to respond to complex environmental concerns in a reliable, adaptable, and evidence-based manner.

For integration to be effective, the models must be in line with: i). national plans for climate action (such as NAPs and NDCs); ii). cycles of regional and urban planning; iii). methods for reducing the risk of disasters (such as the Sendai Framework); and iv). processes related to finance and budgeting (e.g., climate budgeting tools).

5.7.4 Governance and Institutional Capacity

Effective policy integration and decision support for intricate, cross-sectoral issues, such as environmental management and climate change, are made possible in large part by institutional capability and governance. In addition to technical knowledge and resources, strong institutional capacity includes organizational structures, coordination systems, and the capability to process, analyze, and act upon information from a variety of sources. Collaboration between sectors, stakeholders, and government is encouraged through effective governance, which includes participative and polycentric arrangements. This enables policy alignment, adaptive learning, and a prompt response to new threats. According to the literature, even well-designed decision-support tools and evidence-based policies may not be able to influence practice without adequate capacity and coherent governance frameworks. This emphasizes the importance of strengthening institutional resilience, defining roles and responsibilities, and incorporating iterative learning and monitoring mechanisms into policy processes.

- Adaptive modeling flourishes in organizations that facilitate:
  - multidisciplinary cooperation (across industries and departments),
  - free access and transparency of data,
  - Staff and stakeholders should receive training and literacy modeling.
  - Decision-making procedures should be flexible.

Additionally, policy frameworks that are open to new evidence and iterative modifications are required.

5.7.5 Participatory Governance and Co-Production

When models are co-produced with the following, they are more likely to inform actions:

Decision-makers, neighborhood associations, experts and scientists, civil society groups.

This fosters social learning, guarantees relevance, and increases trust, especially when handling contentious or highly valued adaptation choices.

5.7.6 Using Technology to Assist in Decision-Making

Using innovative computational tools, data analytics, and modeling platforms to improve the creation, assessment, and application of policies across sectors is known as "leveraging technology for decision support in policy integration." With the use of technologies such as GIS, Agent-Based Models, System Dynamics Models, and interactive dashboards, policymakers may evaluate trade-offs, simulate scenarios, and predict unforeseen outcomes of interventions in the face of ambiguity. Decision support systems provide evidence-based, transparent, and flexible policymaking by combining real-time data streams, predictive analytics, and visualization tools. In addition to facilitating better coordination across several policy areas, this technological integration makes participatory engagement possible, enabling stakeholders to jointly identify strong solutions and consider alternate approaches. Ultimately, technology-driven decision assistance improves institutions' ability to handle intricately linked social and environmental issues in a changing and unpredictable world.

The usability and accessibility of adaptive modeling outputs are improved by contemporary tools: dashboards on the web with scenario tools and real-time data, decision support systems that are interactive (DSS), mobile applications for farm-level advisories or early warnings, cloud-based systems for sharing data, and collaborating. Urban planning tools, early warning systems, and climate finance tracking can all be connected to these systems.

5.7.7 Learning, Monitoring, and Assessment

By methodically tracking implementation, evaluating results, and iteratively modifying strategies based on evidence, Monitoring, Evaluation, and Learning offers an organized framework for enhancing the efficacy of policies. Decision-makers may integrate cross-sectoral policies, detect unintended impacts, and adjust actions in response to evolving circumstances in complex socio-environmental and climate contexts through Learning, Monitoring, and Assessment (MEL). MEL promotes organizational learning, improves accountability, and supports adaptive governance by fusing quantitative indicators, qualitative evaluations, and feedback systems. By ensuring that policies are dynamic and sensitive to new information, stakeholder insights, and changing SD, their use in decision support improves the robustness and resilience of interventions in the face of uncertainty.

Adaptive policies require ongoing observation and education. The following models are important in:

defining metrics (such as exposure indices and resilience ratings),

Making adjustments to decisions when circumstances or information change establishing feedback channels between science, practice, and policy.

MEL frameworks support long-term resilience and the institutionalization of adaptive activity.

5.8 Conclusion

By specifically addressing uncertainties, system variability, and feedback in socio-ecological and climatic systems, adaptive modeling is essential for improving climate resilience. Through the integration of multi-model ensembles, sensitivity analyses, and decision-support frameworks, such as Robust Decision Making (RDM) and Dynamic Adaptive Policy Pathways (DAPP), adaptive models allow policymakers to identify robust strategies under deep uncertainty and explore a variety of plausible futures. By facilitating monitoring-driven decision updates, early warning systems, and iterative learning, these models guarantee that resilience tactics continue to be successful, even as the effects of climate change vary. The ability of people, ecosystems, and institutions to adapt to unanticipated shocks and long-term climate change is ultimately strengthened by adaptive modeling, which shifts the emphasis from static projections to flexible, evidence-based planning.

Adaptive modeling, which emphasizes the combination of uncertainty quantification, scenario analysis, and iterative learning, should be embraced by practitioners and policymakers as a fundamental element of climate resilience planning. The following are some of the main suggestions: (i) use multi-model ensembles to capture parametric and structural uncertainties; (ii) create monitoring and evaluation systems to prompt timely policy adjustments; (iii) encourage stakeholder engagement and cross-scale coordination to match local actions with more general climate goals; and (iv) institutionalize adaptive governance practices that permit flexibility in management and policy decisions. Decision-makers may create strategies that are strong, adaptable, and able to handle the intricate, nonlinear dynamics of climate change by adopting these methods.

Acknowledgment

The authors express their gratitude to the reviewers and editor for dedicating their valuable time to enhancing this work.

References

Baker, S. R., Bloom, N., Davis, S. J., 2016. Kellogg School of Management, Stanford, & University of Chicago Booth School of Business. *Measuring economic policy uncertainty*. https:// policyuncertainty.com/media/EPU_BBD_Mar2016.pdf

Dawkins, L.C., Bernie, D.J., Lowe, J.A., and Economou, T., 2023. Assessing climate risk using ensembles: A novel framework for applying and extending open-source climate risk assessment platforms. *Climate Risk Management*, 40, 100510, ISSN 2212–0963, https://doi.org/10.1016/j.crm.2023.100510

Haasnoot, M., Fant, V.D., Kwakkel, J., and Lawrence, J., 2024. Lessons from a decade of adaptive pathways research for climate adaptation. *Global Environmental Change*, 88, 102907, ISSN 0959-3780, https://doi.org/10.1016/j.gloenvcha.2024.102907

Herman, J.D., Quinn, J.D., Steinschneider, S., Giuliani, M., and Fletcher, S., 2020. Climate adaptation as a control problem: Review and perspectives on dynamic water resources planning under uncertainty. *Water Resources Research*, 56 (2), e24389.

Janssen, M., Van Der Voort, H., & Wahyudi, A. (2016). Factors influencing big data decision-making quality. *Journal of Business Research, 70*, 338–345. https://doi.org/10.1016/j.jbusres.2016.08.007

Kirschke, S., and Newig, J., 2017. Handling environmental governance and management complexity. *Maintain Science*, 9, 983.

Lehner, F., 2024. Large ensembles of climate models as test sites for applicable compound event studies. *iScience*, 27, 111113. https://doi.org/10.1016/j.isci.2024.111113

Lempert, R.J., Lawrence, J., Kopp, R.E., Haasnoot, M., Reisinger, A., Grubb, M., and Pasqualino, R., 2024. The IPCC's use of decision-making under extreme uncertainty. *Clima Front*, 6, 1380054

Manski, C.F., Sanstad, A.H., and DeCanio, S.J., 2021. Partial identification in climate modeling and policy analysis. *Proceedings of the National Academy of Sciences of the United States of America*, 118 (15), e2022886118. https://doi.org/10.1073/pnas.2022886118,U.S.A

Reis, S., and Nitter, S., 2013. Ecological models, optimization. In Elsevier eBooks. https://doi.org/10.1016/b978-0-12-409548-9.00638-2

Singh, V.P., Maurya, S.P., Singh, R., and Yadav, A.K., 2022. An interdisciplinary modeling approach for dynamic adaptive policy pathways. In S. P. Maurya, A. K. Yadav, & R. Singh (Eds.), *Modeling and Simulation of Environmental Systems: A Computation Approach* (pp. 343–353). CRC Press. https://doi.org/10.1201/9781003203445-22

Stanton, M. C. B., & Roelich, K. (2021). Decision making under deep uncertainties: A review of the applicability of methods in practice. *Technological Forecasting and Social Change*, 171, 120939. https://doi.org/10.1016/j.techfore.2021.120939

Stodden, V., Seiler, J., & Ma, Z. (2018). An empirical analysis of journal policy effectiveness for computational reproducibility. *Proceedings of the National Academy of Sciences, 115*(11), 2584–2589. https://doi.org/10.1073/pnas.1708290115

Uhe, P., Huntingford, C., Mitchell, D., Bates, P., Allen, M., Betts, R., et al., 2021. Reliable confidence assertions of precipitation projections require method-uncertainty. *Journal of Climate*, 34, 11227–1240. 10.1175/JCLI-D-20-0289.

Wainger, L. A., Johnston, R. J., Rose, K. A., Castellini, M. A., McCammon, M., & Newton, J. (2021). Decision Making under Deep Uncertainty. In NOAA SCIENCE ADVISORY BOARD & SAB ECOSYSTEM SCIENCES AND MANAGEMENT WORKING GROUP, *NOAA SCIENCE ADVISORY BOARD*. https://sab.noaa.gov/wp-content/uploads/2021/08/ESMWG_DMDU-Report_Jan2021_Final.pdf

Wu, Y., Miao, C., Fan, X., Gou, J., Zhang, Q., & Zheng, H. (2022). Quantifying the uncertainty sources of future climate projections and narrowing uncertainties with bias correction techniques. *Earth S Future*, *10*(11). https://doi.org/10.1029/2022ef002963

Part II

Climate Change

Towards Sustainable Resource Planning and Management

6

Sustainable Development Goal 13 for Climate Change Actions: Tasks, Methods and Implementation Aspects

Nidhi Gupta, Ravi Kumar, Richa Sharma, Dimpal Chauhan, Naveen Thakur, and Kuldeep Kumar

6.1 Introduction

The atmosphere constitutes a dynamic and perpetually evolving fluid system. Its motion and physicochemical attributes are governed by a complex nexus of interacting forcings. These include solar irradiance, continental geomorphology, oceanographic circulation, orographic configuration and altitude, atmospheric constitution, and terrestrial biomes. The operational timescales of these forcings span several orders of magnitude. For example, alterations in the oceanic thermal distribution, atmospheric chemical composition, and land surface vegetation can manifest over relatively short temporal scales, ranging from diurnal to seasonal variations. Conversely, tectonic processes inducing shifts in continental positioning and orographic relief operate over protracted geological timescales, typically millions of years. The Indian subcontinent experienced a significant warming trend of 0.7°C (1.3 °F) between 1901 and 2018, indicating a discernible shift in its climatic regime. This temperature increase is consistent with global climate change patterns and has contributed to altered weather patterns across the region (Sharma, 2020).

The severe heatwave observed in Pakistan and India during May 2022, with temperatures reaching 51°C, has been attributed to anthropogenic climate change. Climate model simulations suggest that such extreme heat events have become approximately two orders of magnitude more probable because of human influence on the climate system. Specifically, heatwaves exceeding the severity of the 2010 event were previously estimated to have occurred with a return period of 312 years under preindustrial climate conditions. However, current climate projections indicate a significantly reduced return period of approximately 3 years (Rowlatt, 2022). Furthermore, scientific studies utilizing climate models project an increased frequency and intensity

DOI: 10.1201/9781042013234-8

of drought conditions in northern and northwestern India in the near-term. Toward the latter part of the 21st century, a substantial portion of the Indian landmass is anticipated to experience increasingly severe and prolonged drought episodes (Gupta and Jain, 2018).

Changes in precipitation patterns and increased frequency of extreme weather events are also projected to exacerbate the risk of hydrometeorological disasters. For example, states such as Assam are anticipated to witness a rise in the occurrence of severe landslides and floods, posing significant challenges to infrastructure, ecosystems, and human populations. These projections are based on climate model outputs under various emission scenarios, with mid-range scenarios currently considered the most plausible. While India's per capita greenhouse gas emissions remain relatively low relative to those of other major global emitters, its substantial population results in a significant national total. Notably, India's carbon dioxide emissions, although considerably lower than those of China, are exhibiting a more rapid rate of increase. In terms of per capita CO_2 generation, India's contribution is significantly less than that of other primary global regions. In 2023, India experienced a notable increase of 190 million tons in its emissions. This surge was attributed primarily to robust gross domestic product (GDP) growth coupled with a decline in hydroelectricity generation following a weaker-than-average monsoon season. Despite this increase, India's per capita emissions have remained significantly below the global average (Mjeshtri, 2025).

6.2 Threat to Global Health

Climate change poses significant threats to human populations, manifesting as increased flooding, extreme heat events, increased food and water scarcity, disease proliferation, and economic instability (Cifuentes-Faura, 2022; Kumar, et al., 2023). The World Health Organization has identified climate change as one of the preeminent threats to global health in the 21st century. In the absence of mitigation efforts to limit warming, societies and ecosystems will face increasingly severe risks (Willetts, et al., 2022). Adaptation strategies, such as the implementation of flood control measures and the cultivation of drought-resistant crops, can partially mitigate the risks associated with climate change, although inherent limits to adaptation have already been observed (Mukherji, A., 2023; Painter, et al., 2024). Notably, impoverished communities, despite contributing a disproportionately small fraction of global greenhouse gas emissions, possess the least capacity to adapt and exhibit the greatest vulnerability to the adverse impacts of climate change (Tietjen, 2022; Brewer, 2024). Nations with the highest vulnerability to the impacts of climate change have historically contributed minimally to global greenhouse gas emissions. This disparity raises significant questions concerning principles of justice and equity (Fløttum, et al., 2016; Maurya, et al., 2022). Ameliorating global warming significantly facilitates the attainment of the United Nations' Sustainable Development Goals (SDGs), including the eradication of poverty and the reduction of inequalities. This interconnectedness is explicitly acknowledged in SDG 13, which mandates "urgent action to combat climate change and its impacts". Furthermore, goals pertaining to food security, access to clean water, and ecosystem protection exhibit synergistic relationships with climate change mitigation efforts (Roy, et al., 2018).

6.2.1 Interdependence in General

SDG 13 is the cornerstone of all the SDGs. With its emphasis on climate action, SDG 13 is not a stand-alone objective; rather, it is intricately linked to every other SDG. To achieve sustainable development in a number of ways, climate change must be addressed. If left unchecked, climate change multiplies risks and can impede the achievement of the other SDGs. On the other hand, progress toward other SDGs can support efforts to combat climate change. The 2030 Agenda highlights how the SDGs are interrelated and inseparable, necessitating an all-encompassing strategy in which advancements in one objective benefit the others (Figure 6.1).

6.2.1.1 Specific Links to Other SDGs

SDG 1 (No Poverty): Disasters, lower agricultural yields, and health problems are some of the ways in which climate change exacerbates poverty, which disproportionately affects vulnerable communities. Poverty can be decreased by taking climate action, such as encouraging sustainable livelihoods and constructing robust infrastructure.

SDG 2 (Zero Hunger): Food security and agricultural productivity are at risk from droughts, floods, and altered weather patterns caused by climate change. Achieving zero hunger requires climate-resilient farming methods and adaptation plans.

SDG 3 (Good Health and Well-Being): Human health is impacted by climate change, which increases heatwaves, air pollution, and the spread of vector-borne illnesses. Air quality and health consequences are improved by mitigation measures, including supporting sustainable mobility and clean energy.

SDG 4 (Quality Education): Educating people about climate change and enabling them to take action are key components of SDG 4. The incorporation of teaching on climate change into curricula is essential to creating a sustainable future.

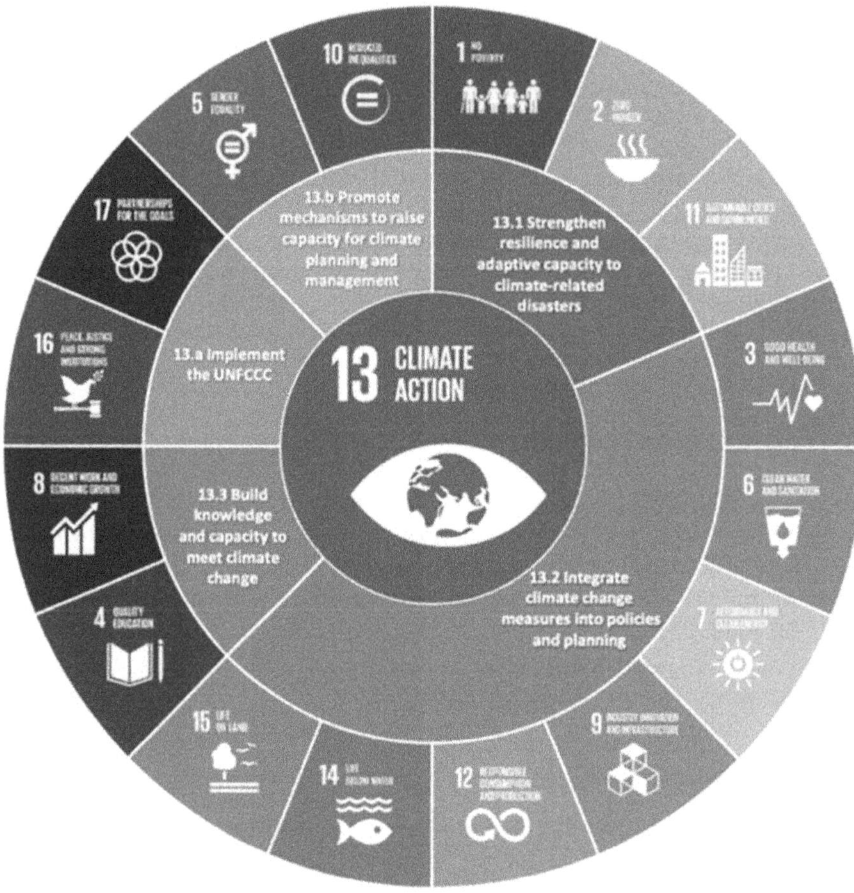

FIGURE 6.1
Relationships among the targets of climate action and the Sustainable Development Goals. ((Scientific Reports | (2023) 13:20582 | https://doi.org/10.1038/s41598-023-47746-w) (Filho, et al., 2023).)

SDG 5 (Equality of Gender): Women are more susceptible to the effects of climate change. Gender-specific vulnerabilities must be taken into account, and women must be empowered to contribute to climate solutions.

SDG 6 (Clean Water and Sanitation): Through changed precipitation patterns and an increase in extreme weather events, climate change has an impact on water availability and quality, according to SDG 6. To provide access to sanitary facilities and clean water, climate adaptation and sustainable water management techniques are essential.

SDG 7 (Affordable and Clean Energy): Making the switch to renewable energy is a crucial mitigation tactic to fight climate change and guarantees that everyone has access to inexpensive, clean energy.

SDG 8 (Decent Employment and Economic Prosperity): As the economy transitions to a green economy, new jobs in energy efficiency, renewable energy, and sustainable industries are created, which promotes both decent employment and economic prosperity.

SDG 9 (Industry, Innovation, and Infrastructure): Reducing emissions and adapting to climate change require fostering sustainable industrialization and innovation as well as the construction of climate-resilient infrastructure.

SDG 10 (Reduced Inequalities): Current inequalities are worsened by the effects of climate change, according to SDG 10. The goal of climate action should be egalitarian, so equitable solutions that safeguard the most vulnerable and advance social justice should be the main emphasis of climate action.

SDG 11 (Sustainable Cities and Communities): Cities are extremely vulnerable to the effects of climate change and contribute significantly to greenhouse gas emissions. It is crucial to create resilient and sustainable cities by implementing strategies such as energy-efficient construction, green infrastructure, and sustainable transportation.

SDG 12 (Responsible Consumption and Production): Reducing resource consumption and greenhouse gas emissions through sustainable patterns of production and consumption helps mitigate climate change.

SDG 14 (Life Below Water): Marine habitats are harmed by ocean warming and acidification caused by rising CO2 emissions. Protecting marine life requires lowering emissions and encouraging ocean protection.

SDG 15 (Life on Land) states that biodiversity loss and climate change are caused by land degradation and deforestation. To combat climate change and preserve terrestrial ecosystems, sustainable land management and forest conservation initiatives are essential.

SDG 16 (Peace, Justice, and Strong Institutions): Resource-related conflicts and displacement may worsen as a result of climate change. Addressing climate-related security threats requires strong institutions and the advancement of justice and peace.

SDG 17 (Partnerships for the Goals): As stated in SDG 17, governments, the commercial sector, civil society, and individuals must work together globally to address climate change.

6.2.2 Agreements and Conferences

In essence, tackling climate change (SDG 13) is an integral part of achieving the entire 2030 Agenda for Sustainable Development. Addressing climate change necessitates immediate and comprehensive interventions. The intrinsic interconnectedness of SDG13 with the broader spectrum of SDGs is legally substantiated through several binding international treaties. These include the United Nations Framework Convention on Climate Change (UNFCCC), the Kyoto Protocol, its subsequent Doha Amendment, and the Paris Agreement, thereby establishing preexisting legal obligations pertaining to SDG13 (Filho, et al., 2022; Kelmam, 2017; Lofs, et al., 2017).

The 2019 Conference of the Parties (COP 25) of the United Nations Framework Convention on Climate Change, held in Madrid, provided a platform for the advancement of critical measures in sectors including finance and agriculture, alongside technology transfer, capacity building initiatives, the rights of indigenous populations, and gender equality. These domains significantly depend on the prevailing climatic conditions. The subsequent Conference of the Parties (COP 26) in Glasgow highlighted a persistent inadequacy in global greenhouse gas (GHG) emission reductions relative to the levels required for maintaining a habitable climate. Furthermore, the findings underscore the insufficient support mechanisms for nations most vulnerable to the adverse impacts of climate change. Nevertheless, COP 26 yielded novel "building blocks" intended to facilitate the implementation of the Paris Agreement, potentially steering the global trajectory toward a more sustainable, low-carbon future (Filho, et al., 2023).

The 27th Conference of the Parties (COP27), held in Sharm el-Sheikh, Egypt, concluded with a pivotal agreement to establish a financial mechanism aimed at supporting countries disproportionately affected by climate-induced disasters such as floods and droughts (Filho, et al., 2023). This outcome was widely acknowledged as a landmark development in global climate governance. However, such progress must be complemented by robust and sustained efforts to reduce greenhouse gas (GHG) emissions resulting from anthropogenic activities. Effective mitigation is essential to minimize the adverse impacts of climate change and strengthen global climate resilience. According to the United Nations Department of Economic and Social Affairs, the global economic slowdown caused by the COVID-19 pandemic led to an estimated 6% decline in GHG emissions in 2020. Nevertheless, this temporary reduction falls significantly short of the sustained annual decreases required to limit the increase in global temperature to 1.5°C above preindustrial levels (Filho et al., 2022; 2023). Although financial commitments to climate action have increased, they continue to lag behind investments directed toward fossil fuel development (Filho et al., 2022; 2023). Moreover, the pace of progress in achieving disaster risk reduction targets remains uneven across countries, underscoring the urgent need for enhanced international cooperation. This aligns with the principles of SDG 17, which emphasizes the importance of global partnerships in advancing sustainable development objectives (Fuso Nerini, et al., 2019; Leal Filho, et al., 2022).

6.3 Tasks and Targets to Combat Climate Change

SDG 13, "Climate Action," calls for urgent action to combat climate change and its impacts. This is a crucial goal given the undeniable threat that climate change poses to the planet and its inhabitants. The specific tasks and targets under SDG 13 are outlined in Table 6.1.

The United Nations Framework Convention on Climate Change (UNFCCC) is the primary international, intergovernmental forum for negotiating the global response to climate change. Source: Adapted from https://sdgs.un.org/goals/goal13

6.3.1 Task-13.1 Strengthening Resilience and Adaptive Capacity to Climate-Related Hazards and Natural Disasters in All Countries

In countries around the world, climate change poses a significant risk to the lives and livelihoods of people. These risks cannot be reduced to zero, which means that governments must take decisive action to help firms and people manage them. Doing so requires planning ahead and putting in place proactive measures that not only reduce climate risk but also accelerate development and reduce poverty, according to a new report *The Adaptation Principles: A Guide for Designing Strategies for Climate Change Adaptation and Resilience.*

The report presents six overarching "Principles of Adaptation and Resilience" and outlines 26 specific actions that governments can implement to formulate effective strategies.

TABLE 6.1

Tasks under Sustainable Development Goal 13

Target	Action
13.1	Strengthening resilience and adaptive capacity
13.2	Integrate climate change measures into national policies, strategies and planning
13.3	Improve education, awareness-raising and human and institutional capacity
13.A	Fully operationalize the Green Climate Fund through its capitalization (USD 100 billion/year)
13.B	Support least developed countries, small island developing States, including focusing on women, youth and local and marginalized communities

Build resilient foundations with rapid and inclusive development
Poverty and a lack of access to basic services including infrastructure, financial services, health care, and social protection are strong predictors of vulnerability to climate change. Effective adaptation strategies critically depend on equipping highly vulnerable populations with essential financial, technical, and institutional resources. Without these crucial provisions, efforts to mitigate the impacts of change invariably fall short, leaving the most susceptible communities unable to cope and adjust.

Help people and firms do their part
It is critical to increase the adaptive capacity of households and firms: many already have incentives to adapt, but they need help overcoming obstacles, ranging from a lack of information and financing to behavioral biases and imperfect markets. Governments can make information on climate risks available, clarify responsibilities and liabilities, support innovation, access the best technologies, and ensure that financing is available to all, especially for solutions that come with high upfront costs.

Revise land use plans and protect critical infrastructure
Governments must safeguard public infrastructure and services alongside direct aid. Developing countries already lose over $390 billion annually from power, water, and transport disruptions. However, with proper data and risk models, making new infrastructure resilient adds only approximately 3% to costs. Public sector urban and land use planning is also crucial, guiding private investments to avoid building in high-risk zones due to long-term climate change.

Help people and firms recover faster and better.
Governments must help people and businesses recover swiftly from disasters, as eliminating all risks is impossible. Prior preparation is key: better weather data, early warning systems, and emergency management significantly reduce damage and losses. For example, advanced hurricane preparation can halve damage. Universal early warning systems offer benefits that are 4-10 times greater in cost. Financial inclusion, such as emergency loans, and social protection are vital for recovery. Adaptive social protection systems, scalable after disasters, are especially efficient but require preestablished delivery and finance mechanisms.

Managing impacts at the macro level
Addressing climate change across all economic sectors demands high-level strategic planning. Climate impacts, from floods affecting housing to ecosystem shifts impacting agriculture, will strain macroeconomic stability and government revenue. Exporting sectors face risks in terms of trade balances and capital flows. Adaptation and resilience spending will further pressure public finances, potentially creating new risks to debt sustainability.

Governments must manage these threats. Given the significant uncertainty in forecasting a climate's macroeconomic impacts, short-term strategies focusing on diversifying economic structure, export composition, and tax bases are particularly beneficial.

Prioritize according to needs, implement across sectors and monitor progress.
Governments need to prioritize adaptation actions on the basis of resources and capacity. They must also create strong institutional and legal frameworks plus a consistent monitoring system. The goal of an adaptation strategy is not a standalone project; it ensures that all government entities integrate and mainstream the strategy into their decisions, continuously monitoring and adjusting actions as needed.

6.3.2 Task-13.2 Integration of Climate Change Measures into National Policies, Strategies and Planning

The Climate Action Tracker (CAT), an independent scientific analysis conducted through a collaboration between Climate Analytics and the New Climate Institute since 2009, quantitatively assesses governmental climate change mitigation policies and commitments. This assessment is benchmarked against the temperature targets outlined in the Paris Agreement, specifically the goal of maintaining global warming well below 2°C and actively pursuing efforts to limit it to 1.5°C.

CAT employs methodologies such as the following:

1. The greenhouse gas (GHG) emission reduction pledges and policy implementation of individual nations should be quantified and evaluated.
2. **Assess the trajectory** of these national actions in relation to their stated commitments and the Paris Agreement goals.
3. **Aggregate** national-level mitigation efforts to project the likely global mean surface temperature increase by the end of the 21st century under different policy scenarios.
4. **Develop sectoral analyses** that delineate potential pathways and associated emission reductions required across various economic sectors to align with global temperature targets.

The five interconnected **domains-People, Planet, Prosperity, Peace, and Partnership**-form the foundational framework for achieving sustainable development. They collectively underscore the comprehensive and integrated approach required to address global challenges and foster a future where human well-being, environmental integrity, economic vibrancy, peaceful coexistence, and collaborative action are mutually reinforced.

6.3.2.1 How do Town Hall COPs Fit into the Current Global Climate Arena?

Town Hall COPs act as crucial local counterparts for global UN climate negotiations. They empower communities to assess their climate progress and develop tailored strategies. Crucially, these localized dialogs directly inform the national climate plans (NDCs) submitted at international forums such as COP30, with the Local Governments and Municipal Authorities (LGMA) constituency facilitating the integration of these local perspectives into national and global climate agendas (Figure 6.2).

Every Town Hall COP is founded upon three core principles:

1. **Town Hall COPs are community-led:**

   COPs are community-driven events that are flexible enough to address local climate needs and priorities. They can either simplify national climate policies for local understanding or highlight existing community climate initiatives to encourage collaboration.

2. **Town Hall COPs are nationally relevant:**

   Insights from community-led Town Hall COPs are nationally important, as new climate plans for the Nationally Determined Contributions (NDCs 3.0) are due at COP30 in 2025. These local discussions can directly inform national climate strategies, support initiatives such as the COP28 Coalition for High Ambition Multilevel Partnership for Climate Action Initiative (CHAMP) and contribute to global climate dialogs. Each Town Hall COP should connect with national climate efforts by inviting government representatives,

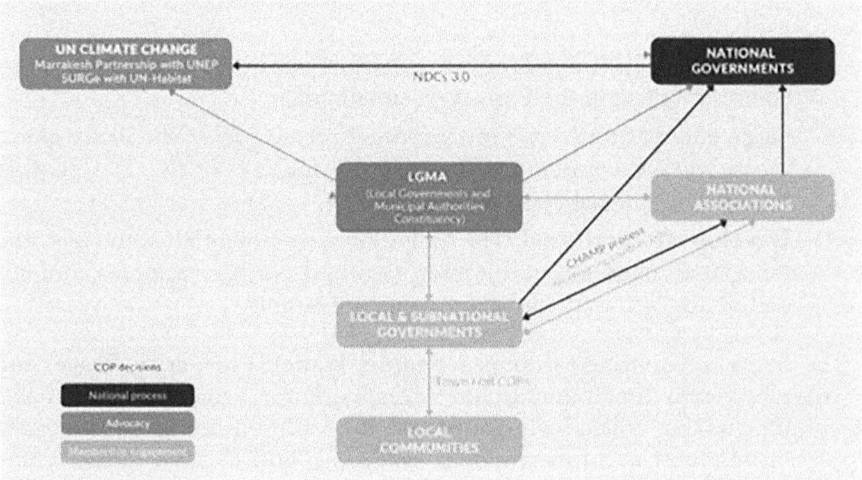

FIGURE 6.2
Interrelations among governments, associations and communities for global climate concerns.

sharing outcomes, and maintaining communication. The ICLEI (Local Governments for Sustainability) will help disseminate these results widely within the UN climate process.

3. **Town Hall COPs are transformation oriented:** These local COPs aim to highlight how climate action directly impacts daily community life, including health, the environment, social justice, affordability, jobs, and cultural heritage. ICLEI's toolkit helps ensure that these events are transformational and relevant to community concerns.

6.3.3 Task-13.3 Improving Education, Raising Awareness and Human and Institutional Capacity

Target 13.3, which focuses on improving climate change education, awareness, and capacity, is crucial for driving effective climate action. Broad public understanding of climate change is a fundamental prerequisite for meaningful engagement. Furthermore, empowering individuals with knowledge and skills, alongside strengthening institutional capabilities, is vital for successfully implementing climate mitigation, adaptation, impact reduction, and early warning measures. The forest sector exemplifies the historical importance of awareness; forest extension programs have long built local capacity for sustainable forest management. Research from Costa Rica, for example, has demonstrated how education and awareness significantly foster tree conservation on agricultural lands (Louman et al., 2016). Strengthening institutional capacity is a core component of REDD+ (reducing emissions from deforestation and forest degradation) and antilegal logging efforts, with nearly $2 billion in recent REDD+ financing dedicated to national capacity building for these initiatives. However, as Chile's experience demonstrates, significant challenges in education, skills development, and institutional capacity persist, hindering successful forest-based climate mitigation and adaptation projects. Overcoming these hurdles is crucial for achieving Targets 13.1 (strengthening climate resilience) and 13.2 (integrating climate measures into national policies). The objective of Target 13.3 is to augment educational, informational, and organizational capabilities concerning climate change mitigation, adaptation strategies, impact attenuation, and early warning systems (Tracker, 2018). The first indicator, 13.3.1, quantifies the degree to which (i) global citizenship education principles and (ii) education for sustainable development methodologies are systematically incorporated into the following components of national education systems:

(a) Formal national education policies and strategic documents.
(b) Prescribed educational curricula across relevant academic levels.
(c) Professional development programs and preservice training for educators.
(d) Standardized student assessment protocols and evaluation methods.

The second indicator indicates the number of sovereign states that have officially reported advancements in reinforcing institutional, systemic, and individual capacities. These reported reinforcements specifically pertain to the operationalization of climate change adaptation measures, greenhouse gas mitigation strategies, and the facilitation of technology transfer and development initiatives related to climate action.

6.3.4 Task-13. A Jointly Mobilizing USD 100 Billion Annually by 2020

Target 13. The SDGs call for developed countries to jointly mobilize USD 100 billion annually by 2020 from a variety of sources to address the needs of developing countries in the context of meaningful mitigation actions and transparency in implementation and to operationalize the Green Climate Fund through its capitalization as soon as possible.

At the time of this analysis, USD 10.3 billion had been pledged to the Green Climate Fund (GCF). However, the estimated overall costs of climate change actions are significantly greater, with projections reaching approximately USD 2.4 trillion for the energy sector alone (De Coninck et al., 2018). Compared with certain climate actions within the energy, transport, and industrial sectors, investments in the forest sector present a relatively cost-effective avenue for contributing to the overarching goals of reducing atmospheric carbon concentrations and enhancing resilience, particularly for communities that depend on forests for their livelihoods.

An examination of past allocations of climate funds toward forests (Bird et al., 2017) suggests that if Target 13. A were to be fully met, approximately USD 3 billion per year could become available for emission reduction initiatives within the forest sector by 2020. Theoretically, this annual sum of USD 3 billion could cover approximately 12% of global forest emissions. While substantial REDD+ funding, nearly $2 billion, has been directed toward strengthening national capacities for forest-based climate action, further analysis is crucial to understand its effective utilization. A significant portion of these funds will likely need to establish foundational conditions for measurable emission reductions, which may take considerable time for many nations to achieve. Additionally, current climate fund policies require supporting forest landscape restoration (FLR) efforts. However, current funding levels are clearly insufficient to fully unlock the vast mitigation and adaptation potential that forests offer.

6.3.5 Task-13. B Promotion of Mechanisms to Raise Capacity for Planning and Management

Target 13.b of the SDGs aims to enhance the capabilities of least developed countries (LDCs) and small island developing states (SIDS) for efficacious climate change adaptation and mitigation planning and implementation.

This includes a specific focus on empowering women, youth, and local and marginalized communities as critical stakeholders and agents of change. The target explicitly recognizes the United Nations Framework Convention on Climate Change (UNFCCC) as the central multilateral platform for international climate negotiations and the establishment of global climate action frameworks.

Specifically, Indicator 13.b.1 serves as a quantitative metric to assess the progress of Target 13.b. It tracks the number of least developed countries (LDCs) and small island developing states (SIDS) that have formulated and officially communicated key strategic climate policy documents. These documents include:

- **Nationally Determined Contributions (NDCs):** These outline a country's self-determined goals for reducing greenhouse gas emissions and adapting to the impacts of climate change, as per the Paris Agreement under the UNFCCC. They represent a nation's commitment to global climate action.
- **Long-term Strategies (LTS):** These strategic documents articulate a country's long-term vision for transitioning to a low-emission and climate-resilient development pathway, often looking toward the middle century or beyond.
- **National Adaptation Plans (NAPs):** These plans identify a country's medium- to long-term adaptation needs and strategies, aiming to reduce vulnerability to the adverse effects of climate change and integrate adaptation into national development planning.
- **Strategies as reported in Adaptation Communications:** Adaptation Communications are submissions to the UNFCCC through which countries can share their adaptation priorities, plans, actions, and support needs. Reporting strategies within these communications signifies a proactive approach to adaptation planning.
- **National Communications (NCs):** These are reports submitted by parties to the UNFCCC, providing comprehensive information on their greenhouse gas emissions, policies and measures to address climate change, and adaptation efforts.

By monitoring the number of LDCs and SIDSs that have submitted these reports, Indicator 13.b.1 provides a measure of the extent to which these vulnerable nations are actively engaging in climate-related planning and developing institutional and strategic frameworks for effective management. An increase in this indicator over time suggests enhanced capacity and commitment within these countries to address climate change challenges. These data can inform international efforts to provide targeted support and resources to further build their capacity in this critical area.

6.4 Methodology: Exploratory Study on the Influence of Sustainable Development Goal 13

To investigate the impact of SDG 13 (climate action) on the implementation of other SDGs, an international exploration study was conducted. The European School of Sustainability Science and Research (ESSSR) and the Inter-University Sustainable Development Research Programme (IUSDRP) distributed a survey to their respective networks. Expert respondents were requested to provide answers to three specific inquiries:

(i) their country of affiliation (closed-ended question with a predefined list of countries);
(ii) their primary professional role (closed-ended question with the options: Research, Teaching, Administration, and Other); and
(iii) "To what extent does the implementation of SDG13 (climate action) positively influence the implementation of the other SDGs?"

For the third question, a Likert-type scale was used to gauge the perceived positive influence of SDG13 implementation on the other SDGs (SDGs 1–12 and 14–17). The response options provided were no influence, a little influence, moderate influence, to some extent, and to a great extent.

A broad invitation to participate in the survey was disseminated throughout the networks of the ESSSR and IUSDRP, which targeted researchers, teaching and administrative personnel, and other representatives from universities spanning all geographical regions. Recipients were also encouraged to extend the invitation to their respective institutional networks to maximize dissemination. Given the voluntary nature of participation, the authors exerted no control over individual respondents' decisions to engage with the survey. The only explicit prerequisite for participation in the survey was a demonstrable familiarity with the central research topics: climate action and its interdependencies with the remaining SDG.

6.5 Links between Sustainable Development Goal 13 and Food Systems

The connection between climate action and food systems is profound and operates in two main directions:

6.5.1 Food Systems as Drivers of Climate Change

Greenhouse gas emissions: Agriculture, including livestock, crop production, and land-use change (such as deforestation for agriculture), is a significant contributor to greenhouse gas emissions (methane, nitrous oxide, and carbon dioxide).

Resource depletion: Unsustainable agricultural practices can lead to soil degradation, water scarcity, and biodiversity loss, further exacerbating environmental challenges.

Food Waste: Significant amounts of food are lost or wasted throughout the supply chain, contributing to unnecessary emissions from production, transportation, and decomposition.

6.5.2 Climate Change as a Threat to Food Systems

Extreme Weather Events: The increased frequency and intensity of droughts, floods, heatwaves, and storms disrupt agricultural production, damage infrastructure, and impact food security.

Shifts in growing seasons and crop suitability: Changes in temperature and precipitation patterns can alter optimal growing conditions for crops and livestock, leading to reduced yields and potential food shortages.

Increased Pests and Diseases: Climate change can create favorable conditions for the spread of agricultural pests and diseases, further threatening crop and livestock production.

Sea Level Rise and Coastal Erosion: These phenomena can impact coastal agricultural lands and aquaculture.

Climate change manifests through increased extreme weather events, including more frequent and intense droughts, floods, heatwaves, and storms, all of which severely disrupt agricultural production, damage critical infrastructure, and consequently jeopardize global food security. Furthermore, altered temperature and precipitation patterns are shifting the optimal growing season and crop suitability, leading to reduced yields and potential food shortages. These climatic changes also foster the proliferation of agricultural pests and diseases, posing an additional threat to crop and livestock production. In combination with these issues, rising sea levels and coastal erosion directly impact agricultural lands and aquaculture in vulnerable coastal regions. To address climate change effectively, national policies under SDG 13 must explicitly integrate food systems, acknowledging their contributions to and vulnerability to climate change. This involves developing specific mitigation and adaptation strategies for food systems, such as promoting climate-smart agriculture, reducing food loss and waste, and building resilient supply chains. In addition to process indicators, it is beneficial to establish outcome-oriented indicators to measure tangible impacts, such as reduced agricultural emissions or increased food production resilience. Finally, leveraging synergies with other SDGs, such as Zero Hunger (SDG 2), Life on Land (SDG 15), and Responsible Consumption and Production (SDG 12), is crucial for a comprehensive and effective approach.

6.6 Implementation Aspects

A scientifically rigorous approach to SDG 13 necessitates the consideration of its systemic interdependencies with other SDGs and the institutional landscape governing their implementation. Climate change, the focus of SDG 13, functions as a critical and dynamic forcing factors influencing the long-term viability of food systems and the environmental parameters upon which they depend, including water resources, terrestrial ecosystems, oceanic health, and the frequency and intensity of natural hazards (Vermeulen, et al., 2012; Campbell, et al., 2018). Consequently, effective climate action must be conceived and executed within a framework that explicitly acknowledges and leverages these intricate cross-sectoral linkages to foster synergistic outcomes and avoid unintended negative consequences across the broader sustainable development agenda.

The bidirectional relationship between climate change and food systems is characterized by intricate feedback loops. Climatic impacts on food systems cascade across multiple dimensions of sustainable development, significantly affecting poverty levels, public health outcomes, economic stability, infrastructural integrity, societal equity, and gender dynamics (Vermeulen, et al., 2012). Conversely, food production practices, alongside energy generation and unsustainable consumption and production patterns, contribute substantially to greenhouse gas emissions, thereby exacerbating climate change. From a developmental standpoint, realizing effective adaptation and mitigation strategies within food systems is contingent upon progress in several enabling SDGs that underpin SDG 13. These include fostering sustainable production and consumption (SDG 12), ensuring food security (SDG 2), alleviating poverty (SDG 1), promoting quality education (SDG 4), achieving gender equality (SDG 5), ensuring access to clean water and sanitation (SDG 6), protecting terrestrial ecosystems (SDG 15), and transitioning to affordable and clean energy (SDG 7). The specific manifestations and strengths of these interdependencies are further modulated by geographical, technological, and governance contexts (Nilsson, et al., 2016). Between 15 and 30% less emissions may be produced if food supply chain losses are reduced to promote sustainable production and consumption (Nash, et al., 2017).

There are significant connections between food security, poverty, equity, and adaptation in SDG 13. Increased synergies between SDG 13 mitigation and agricultural energy, water and nutrient input efficiency are also anticipated (Figure 6.3).

6.6.1 Role of Space-Based Technologies in Addressing Climate Change and Disaster Risk Reduction, Aligning Directly with the Aims of SDG 13

Space-based technologies play a crucial role in addressing climate change and disaster risk reduction, directly supporting the objectives of SDG 13. Initiatives such as UNOOSA's GP-STAR demonstrate the application of space technology, including Earth observation, GNSS, and satellite telecommunications, to enhance disaster risk reduction, which aligns with the Sendai Framework. The momentum generated by post-COP21 efforts and the One Planet Summit underscores a global commitment to leveraging space-based resources in climate action. The proposed Space Climate Observatory (SCO) is poised to

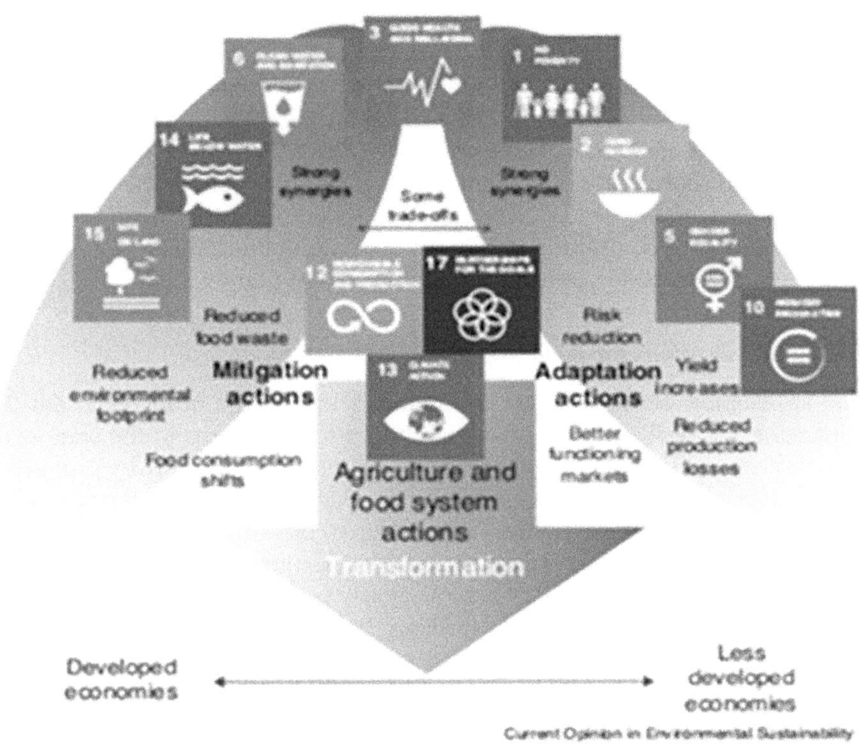

FIGURE 6.3
Relationships between climate change actions in the food system and sustainable development goals. (Current Opinion in Environmental Sustainability 2018, 34:13–20 (Campbell, et al., 2018).)

significantly advance climate understanding through enhanced, sustainable, and accessible satellite-derived data, fostering global collaboration among space agencies and stakeholders. By promoting open data policies, the SCO aims to broaden data utilization for research, policy formulation, and public awareness, contributing strategically to the Space 2030 Agenda and facilitating the development of low-emission, climate-resilient societies.

6.6.1.1 Link to SDG 13

The aforementioned initiatives directly support several facets of SDG 13. Specifically, GP-STAR's emphasis on disaster risk reduction strengthens resilience and adaptive capacity to address climate-related hazards and natural disasters (Target 13.1) through tools that facilitate early warning, risk assessment, and disaster response, leveraging Earth observation data for monitoring extreme weather events. Furthermore, the data and insights derived from initiatives such as the SCO inform national climate policies and adaptation strategies, enabling more evidence-based decision-making and the integration of climate change measures into national policies, strategies, and planning (Target 13.2). By promoting open access to climate data and fostering collaboration, these initiatives also enhance education, awareness-raising, and human and institutional capacity for climate change mitigation, adaptation, impact reduction, and early warning (Target 13.3). While not explicitly articulated, the capacity of GP-STAR and the SCO to provide accessible data and tools holds the potential to significantly benefit the developing countries and small island developing states, including their women, youth, and marginalized communities, thereby promoting mechanisms for effective climate change-related planning and management (Target 13.B).

6.6.1.2 Implications for Food Systems

The application of space-based technologies offers significant potential for enhancing the resilience and sustainability of food systems in the context of climate change. Satellite imagery facilitates precision agriculture by providing critical data on crop health, soil moisture, and land use, enabling the optimization of agricultural practices to reduce resource consumption and greenhouse gas emissions. Earth observation capabilities also improve drought and flood monitoring, enabling the development of early warning systems that enhance preparedness and mitigate food shortages. Furthermore, satellites play a crucial role in tracking deforestation rates, which is essential for addressing a key driver of climate change and its adverse effects on agricultural land. Enhanced climate data from initiatives such as the SCO improve the accuracy of climate models, generating more reliable long-term forecasts for agricultural planning and adaptation. Finally, satellite imagery supports disaster response in agricultural regions by providing critical information for damage assessment and facilitating targeted relief efforts following climate-related events.

6.7 Strengths of Sustainable Development Goal 13

Dual Focus: By addressing both adaptation (strengthening resilience, integrating climate measures) and mitigation (although lacking a specific target, the overall aim implies it), SDG 13 acknowledges the multifaceted nature of climate change.

Emphasis on Integration: The call to integrate climate change measures into national policies and planning is a significant strength. This promotes a mainstream approach, ensuring that climate action is considered across various sectors, including potentially food systems.

Focus on Enabling Factors: The goal is that achieving climate action requires robust processes, monitoring of financial commitments, and improved capacity, particularly in vulnerable nations and communities. This focus on "how" is crucial for effective implementation.

Equity Considerations: Explicitly mentioning the need to improve capacity among LDCs, SIDS, women, youth, and marginalized communities highlights the importance of climate justice and inclusive action.

6.7.1 Limitations of SDG 13

Process-Oriented Indicators: A key limitation is that SDG 13 largely covers processes rather than outcomes. While processes are important, the lack of specific outcome-based indicators can make it challenging to directly measure the impact of climate action.

Absence of a Mitigation Target: The lack of a concrete, measurable mitigation target within SDG 13 is a significant gap. Mitigation is crucial for limiting the extent of future climate change, and a specific target would provide a clearer benchmark for progress.

Indirect Link to Food Systems: While the need to integrate climate measures into national policies could include food systems, the direct link is not as explicit as in some other SDGs (e.g., SDG 2 on Zero Hunger). This might lead to less focused attention being given to the intricate relationship between climate change and food security.

6.8 Conclusion

Achieving SDG 13 for climate action is not merely an environmental imperative but a foundational pillar for holistic sustainable development. Its tasks, methods, and implementation underscore the urgent need for a multifaceted and integrated approach. From strengthening resilience and adaptive capacity at the local and national levels to mainstream climate considerations across all sectors and, critically, enhancing education, awareness, and institutional capacity, SDG 13 demands concerted global effort. The effectiveness of these actions hinges on robust institutional frameworks, strategic resource allocation, and a commitment to addressing the systemic inequalities that exacerbate climate vulnerability. Ultimately, by prioritizing early interventions, investing in vulnerable populations, and fostering comprehensive, data-driven strategies, we can accelerate progress toward a more resilient, equitable, and sustainable future for all, ensuring that climate action is not just a goal but an ingrained practice in every facet of governance and society.

Acknowledgment

The authors sincerely thank their respective prestigious institutes and others who were involved in this study for their direct and indirect contributions.

References

Bird, J., Murphy, R., Ciappio, E., & McBurney, M. (2017). Risk of deficiency in multiple concurrent micronutrients in children and adults in the United States. *Nutrients, 9*(7), 655. https://doi.org/10.3390/nu9070655

Brewer, T., 2024. *Climate Change: An Interdisciplinary Introduction*. Springer Nature.

Campbell, B.M., Hansen, J., Rioux, J., Stirling, C.M. and Twomlow, S., 2018. Urgent action to combat climate change and its impacts (SDG 13): transforming agriculture and food systems. *Current Opinion in Environmental Sustainability, 34*, pp.13–20. https://doi.org/10.1016/j.cosust.2018.06.005

Cifuentes-Faura, J., 2022. European Union policies and their role in combating climate change over the years. *Air Quality Atmosphere & Health, 15*(8), pp.1333–1340. https://doi.org/10.1007/s11869-022-01156-5

Cooper, J., Dubey, L. and Hawkes, A., 2022. The life cycle environmental impacts of negative emission technologies in North America. *Sustainable Production and Consumption, 32*, pp.880–894.

De Coninck, H., Ph. D., Revi, A., Babiker, M., Bertoldi, P., Buckeridge, M., Cartwright, A., Dong, W., Ford, J., Fuss, S., Hourcade, J.-C., Ley, D., Mechler, R., Newman, P., Revokatova, A., Schultz, S., Steg, L., & Sugiyama, T. (2018). Strengthening and implementing the global response. In *Global Warming of 1.5°C. An IPCC Special Report on the impacts of global warming of 1.5°C above pre-industrial levels and related global greenhouse gas emission pathways, in the context of strengthening the global response to the threat of climate change, sustainable development, and efforts to eradicate poverty* (pp. 313–444). Cambridge University Press. www.ipcc.ch/site/assets/uploads/sites/2/2022/06/SR15_Chapter_4_LR.pdf

Filho, W.L., Hickmann, T., Nagy, G.J., Pinho, P., Sharifi, A., Minhas, A., Islam, M.R., Djalanti, R., García Vinuesa, A. and Abubakar, I.R., 2022. The influence of the corona virus pandemic on sustainable development goal 13 and United Nations framework convention on climate change processes. *Frontiers in Environmental Science, 10*, p.784466.

Filho, W.L., Wall, T., Salvia, A.L., Dinis, M.A.P. and Mifsud, M., 2023. The central role of climate action in achieving the United Nations' sustainable development goals. *Scientific Reports, 13*(1), p.20582.

Fløttum, K., Gasper, D. and Clair, A.L.S., 2016. Synthesizing a policy-relevant perspective from the three IPCC "Worlds"-A comparison of topics and frames in the SPMs of the Fifth Assessment Report. *Global Environmental Change, 38*, pp.118–129.

Fuso Nerini, F., Sovacool, B., Hughes, N., Cozzi, L., Cosgrave, E., Howells, M., Tavoni, M., Tomei, J., Zerriffi, H. and Milligan, B., 2019. Connecting climate action with other sustainable development goals. *Nature Sustainability, 2*(8), pp.674–680.

Gupta, V. and Jain, M. K., 2018. Investigation of multimodel spatiotemporal mesoscale drought projections over India under climate change scenario. *Journal of Hydrology, 567*, 489–509.

Kelman, I., 2017. Linking disaster risk reduction, climate change, and the sustainable development goals. *Disaster Prevention and Management: An International Journal, 26*(3), pp.254–258.

Kumar, A., Chava, R., Gupta, S., Shirin, S., Jamal, A. and Yadav, A.K., 2023. Environmental and health effects of fluoride contamination and treatment of wastewater using various technologies. In A. K. Yadav, S. Saba, & V. P. Singh (Eds.), *Advanced Treatment Technologies for Fluoride Removal in Water* (pp. 323–341). Springer Nature Switzerland. https://doi.org/10.1007/978-3-031-38845-3_18

Leal Filho, W., Wall, T., Barbir, J., Alverio, G.N., Dinis, M.A.P. and Ramirez, J., 2022. Relevance of international partnerships in the implementation of the UN sustainable development goals. *Nature Communications, 13*(1), p.613.

Lofts, K., Shamin, S., Zaman, S.T. and Kibugi, R., 2017. Brief on sustainable development goal 13 on taking action on climate change and its impacts: contributions of international law, policy and governance. *McGill Journal of Sustainable Development Law, 13*, p.183.

Louman, B., Gutiérrez, I., Le Coq, J.F., Brenes, C., Wulfhorst, J.D., Casanoves, F., Yglesias, M. and Rios, S., 2016. Avances en la comprensión de la transición forestal en fincas costarricenses. *Revibec: revista iberoamericana de economía ecológica, 26*, pp.191–206.

Maurya, S.P., Yadav, A.K. and Singh, R. (Eds.), 2022. *Modeling and Simulation of Environmental Systems: A Computation Approach* (1st ed.). CRC Press. https://doi.org/10.1201/9781003203445.

Mjeshtri, J., 2025. *Utilization of the Household Carbon Footprint Assessment for Equitable Policy Formulation in Greece* (Master's thesis).

Mukherji, A., 2023. Climate change 2023 synthesis report. IPCC. (2023). *Climate Change 2023: Synthesis report* (Core Writing Team, H. Lee, & J. Romero, Eds.; pp. 35–115). https://doi.org/10.59327/IPCC/AR6-9789291691647

Nash, J., Peña, O., Galford, G.L., Gurwick, N., Pirolli, G., White, J.M. and Wollenberg, E.K., 2017. Reducing food loss in agricultural development projects through value chain efficiency. CCAFS Working Paper no. 204. Wageningen, The Netherlands: CGIAR Research Program on Climate Change, Agriculture and Food Security (CCAFS).

Nilsson, M., Griggs, D. and Visbeck, M., 2016. Policy: map the interactions between sustainable development goals. *Nature, 534*(7607), pp.320–322.

Painter, J., Marshall, S. and Leitzell, K., 2024. Communicating climate futures: a multicountry study of how the media portray the IPCC scenarios in the 2021/2 Working Group reports. *Climatic Change, 177*(6), p.82.

Rowlatt, J., 2022. Climate change swells odds of record India, Pakistan heatwaves. *BBC News*, 18.

Roy, J., Tschakert, P., Waisman, H., Abdul Halim, S., Antwi-Agyei, P., Dasgupta, P., Hayward, B., Kanninen, M., Liverman, D., Okereke, C. and Fernanda Pinho, P., 2018. Special report on global warming of 1.5C (SR15)-Chapter 5: sustainable development, poverty eradication and reducing inequalities. In Intergovernmental Panel on Climate Change (Ed.), *Global Warming of 1.5°C. An IPCC Special Report on the impacts of global warming of 1.5°C above pre-industrial levels and related global greenhouse gas emission pathways, in the context of strengthening the global response to the threat of climate change, sustainable development, and efforts to eradicate poverty* (pp. 445–538). Cambridge University Press. www.ipcc.ch/site/assets/uploads/sites/2/2022/06/SR15_Chapter_5_LR.pdf

Sharma, V., 2020. Average temperature over India projected to rise by 4.4 degrees celsius: govt report on impact of climate change in country. Tribune India.

Tietjen, A. K., Kroemer, N., Cattaneo, D., Baldelli, S., & Wicha, S. G. (2021). Population pharmacokinetics and target attainment analysis of linezolid in multidrug-resistant tuberculosis patients. *British Journal of Clinical Pharmacology, 88*(4), 1835–1844. https://doi.org/10.1111/bcp.15102

Tracker, S.D.G., 2018. Measuring progress toward the sustainable development goals. *SDG-Tracker*. Available online: https://sdg-tracker.org/ (accessed on 27 December 2021).

Vermeulen, S.J., Campbell, B.M. and Ingram, J.S., 2012. Climate change and food systems. *Annual Review of Environment and Resources, 37*(1), pp.195–222.

Willetts, E., Campbell-Lendrum, D., World Health Organization, & Intergovernmental Panel on Climate Change. (2022). Review of IPCC Evidence 2022: Climate Change, Health, and Well-being. In *Policy Brief.* https://cdn.who.int/media/docs/default-source/climate-change/who-review-of-ipcc-evidence-2022-adv-version.pdf?sfvrsn=cce71a2c_3&download=true

7

Impact on Climate Change Due to Land Use and Land Cover: Assessment of Vishakhapatnam District, Case Study

N.G. Kalyani Reddy, Peddada Jagadeeswara rao, and Victor Babu Neela

7.1 Introduction

Changes in land use and land cover (LULC) are important aspects of global ecological change. Satellite remote sensing (RS) data, which are plentiful, have proven useful in charting LULC trends over time. The typical economic necessities of humanity in the real world determine a region's LULC. Owing to severe rural and segment demand, land is becoming a limited resource. LU and LC classification via RS images is currently crucial for many applications, including land-use planning, agricultural operations, biological resource management, and forest management. Understanding LULC data is essential for planning for the future on the basis of climate change, ensuring the viability of available resources, and comprehending the land's existing conditions. Urban sprawl, forest cover, agricultural land, waste, water, open land, and industrial land all contribute to the protection of the Earth's surface. The way that people use topography for sustainability and improvement is revealed by land use. In the future, environmental conditions will be impacted by human activity as the population grows tremendously. Rapid population expansion, inadvertent decolonization, and sprawl all seem to have disastrous repercussions on the surrounding territory. To resolve environmental conflicts, including unstructured evolution; the loss of agricultural land; and the devastation of nature at the territorial, national, and international scales, LULC evaluation is crucial.

LULC modification surveys aid in the explanation of the following: (i) the type of land cover that is replacing the previous type, (ii) the consequences occurring, (iii) the rate of land change, (iv) the ultimate cause of the change, and (v) the type of alteration occurring. The rapid globalization of administrative divisions has become an environmental issue in the twenty-first century, one that calls for the adoption of new scientific theories and data

DOI: 10.1201/9781042013234-9

sources. These abrupt changes in the local, territorial, and global environments have served as serious warnings to people. For environmental effect analysis, the desired information is thus provided by a collection of data from RS and geographic information system (GIS) applications with selected spatial and spectral resolutions.

The phenomenon of RS is used to observe or gather data about far-off objects without making physical contact, in contrast to observations in situ or onsite. The science and art of determining current information about an object, region, or process by analyzing data collected by a device that is not in close proximity to the object, region, or process being studied is known as RS. Specifically, RS is defined as the accumulation and visualization of data about a physical object or region without having direct physical contact with the object or region. RS enables the monitoring of environmental changes and can be used to assess the state of a planet over a very wide area.

Using the DEM, the property bent surface region of the articles was extracted, and their three-layered powerful change data which recognized four-layered ranger service data observations on the schedule and geographical levels were broken down. Formal processes and mapping are carried out via field research, maps, and accessible records (Ademiluyi, et al., 2008). Formal processes are therefore expensive and time-consuming. Additionally, noncurrent variables in the dynamic environment are swiftly transformed by the maps that are produced. Aerial photographs and high-resolution images can be used to obtain LULC changes in large cities; however, because of the presence of commercial enterprise elements, the database is limited. Nonetheless, medium-resolution data for LULC change detection are accessible worldwide and include Landsat5, Landsat8 OLI pictures, multispectral scanner (MSS), thematic mapper, and ETM+ images.

Although counting samples takes a lot of time, more samples are required for good precision. Therefore, accuracy and time complexity are traded off. Accuracy is increased at the expense of time complexity if the number of samples increases (Sajib, et al., 2022; Rastogi, et al., 2025). The maximum likelihood (ML) classifier has low accuracy. To validate the evaluation of the population's variance–covariance matrix and mean vector, sufficient ground truth data should be sampled. Therefore, when a signature file is used, the interactive supervised classification technique is utilized to highlight the supervised class to address poor accuracy issues.

7.2 Objective

Many improvements are being made to visit Visakhapatnam, a popular tourist destination in the state of Andhra Pradesh, to increase its economic standing. The primary goal in this regard is (i) to examine LULC changes in Visakhapatnam and evaluate spatiotemporal variations in light of LULC changes. (ii) To examine how classes have changed between vegetative and urbanized areas (Kathwas, et al., 2019), data from Open Street Map, Google Earth Pro, and Bhutan were compared over a 15-year period. The study illustrates how various groups relate to one another during a period of rapid urbanization and how changes in land cover and use might impact this relationship. Study area Andhra Pradesh's most important and populated town. It is situated between the coastline of the Bay of Bengal and the Eastern Ghats. According to the data, it is the fourth-largest town in South India and the second-largest town on the east coast of India after Chennai. According to the newspaper of India, it is one of Andhra Pradesh's four smart cities designated under the Smart Cities Mission. The town is situated between 17.7041 N and 83.2977 E. The climate in Visakhapatnam is arid, swampy, and equatorial.

7.3 Methodology

The main focus of this investigation was to combine statistical data and satellite imagery (Kotoky, et al., 2012) to illustrate the effects of land use. In this study, the numerical processes of alternate detection are utilized. Every satellite image is classified via the alternative detection technique. Using a change detection matrix, the resulting LULC maps obtained after the type are compared to the pixel-by-pixel conception.

This investigation focused primarily on the use of statistical data and satellite imagery to illustrate the effects of land use. In this study, the numerical processes of alternate detection are utilized. Every satellite image is classified via the alternative detection technique. Using a change detection matrix, the resulting LULC maps obtained after the type are compared to the pixel-by-pixel conception, as shown in Figure 7.1.

Step 1: Data Gathering: LULC maps of Vishakhapatnam for the years 2015, 2021, and 2025 were created using data from Landsat satellites (Thematic Mapper TM) and the Operational Land Imager OLI) (OLI 9 February). OLI on December 28, 2021, and OLI on February 19, 2025. The flowchart of the suggested method with a 30 m resolution. The spatial data of the study area are shown in Table 7.1. The global framework for Landsat (Kanmani, et al., 2023) information documentation is called the Worldwide Reference System (WRS). Through the determination of an apparent aspect of place assigned by path and row numbers, WRS enables a consumer to inquire about satellite pictures of any region of the universe. Unambiguously, a given aspect center was confirmed by the accumulation of a path and row number. The first is the path number, which is always followed by the row number. USGS Earth Explorer was used to download the pictures.

TABLE 7.1

Spatial information of the study area

Component	Information
AREA	VISHAKAPATNAM
DATA SOURCE	LANDSAT
SENSOR	ETM+, OLI/TIRS
PATH	141
ROW	48

Step 2: Image preprocessing: An inaccuracy that affects a scene element's radiometric value or brightness is called radiometric rectification. To standardize images and increase interpretability and analysis, radiometric correction is used to reduce the number of mistakes in digital images. Radiometric and geometric corrections, which are acquired via picture preprocessing techniques, are essential for collecting different types of original data (Paul, et al., 2023) in a categorization process. Layer stacking and the area of interest were used to merge several bands into a single image. To improve pixel clarity, edge enhancement concurrently carries out brightness, contrast, and noise cancellation.

Step 3: Supervised classification: Training samples for classification are obtained via Bhuvan source sampling. The process of sampling entails breaking the aggregation into smaller subgroups called strata. The strata in stratified random sampling are created according to certain qualities or traits. Data classes are extracted from a multiband raster image via image classification. It falls into one of two categories: uncontrolled or monitored. To extract data categories from a multiband raster image, images were classed. This study suggested a novel method that reveals a supervised class by employing source image samples for classification.

It speeds up the classification process. All of the bands in the chosen image with built-in pyramids are used in supervised categorization. For better usability, it takes advantage of the image's true pyramid-level resolution. Raster datasets display better results when pyramids are constructed. Polygons enclosing particular land cover features that were representative of the required land cover categories were digitized to create training sites.

To give each desired cover type a single class ring from numerous training sites, the training sites from each group were then combined into a single class. A pixel with the highest likelihood was placed in the proportionate class after the entire survey field was divided into six land-use groups: urban sprawl, forest cover, agricultural land, waste, water, open land, and industrial land.

Step 4: Confusion matrix production and LULC change detection: A transition matrix is produced and used to evaluate RS photos. The accurately categorized elements, which determine the classification accuracy, are

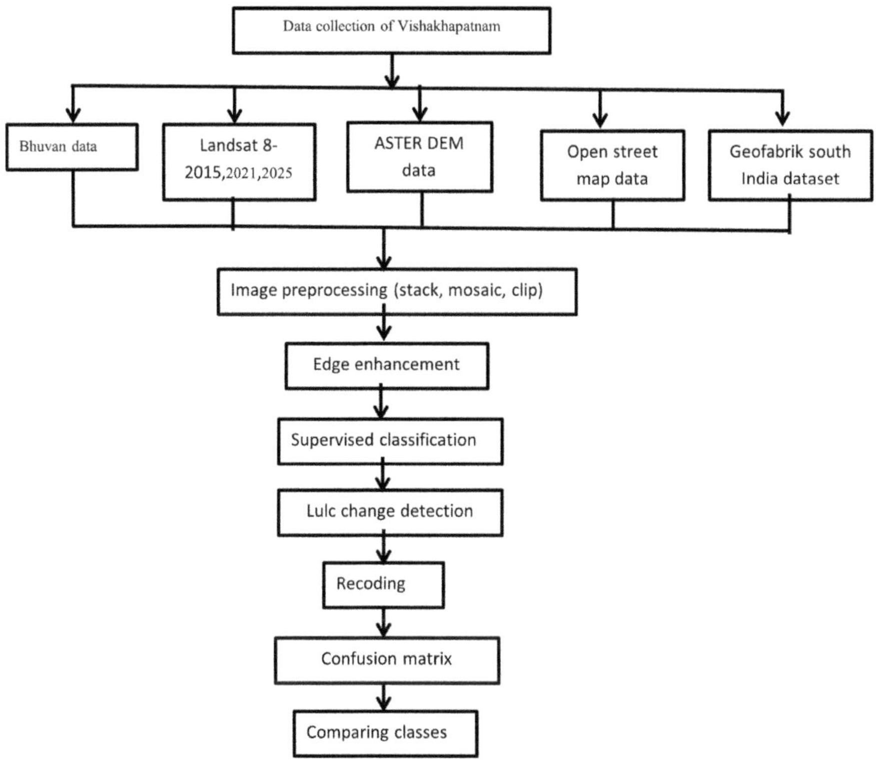

FIGURE 7.1
Methodology for land use and land cover classification.

represented by the diagonal members of the matrix. The transition matrix shows multiple mistakes in the classification process, enabling more accurate assessment and strengthened map evaluation.

Step 5: Assessment of accuracy: The overall accuracy, or the ratio of the total number of correctly identified pixels to the total number of pixels, was then computed for the classified theme output.

7.4 Results

The survey region's land use and land cover (LULC) is divided into built-up, forest, vegetative agriculture, open land, water bodies, and industry. The classified photos of the three regions in 2015, 2021, and 2025 are shown in Figures 7.2--7.4. The survey region's LULC effects are shown in Table 7.2. The error matrix illustrates how the built-up area increased (Marić, et al., 2022) between 2015 and 2015. Throughout the study period, there was an exponential increase in the total built-up area. Between 2015 and 2021, the percentage of forest increased from 27.865% to 27.860%. Between 2015 and 2021, the share of agriculture increased from 42.13% to 41.96%, as shown in Figure 7.2.

A critical phase in RS classification is accuracy assessment, which determines how closely a classified image resembles reality. It aids in error detection and assessing the accuracy of land cover classification. A confusion matrix that compares categorized picture pixels with reference data is used in ERDAS Imagine to evaluate accuracy. Because of its geospatial accuracy and high-resolution imagery, Google Earth Pro is frequently used as a reference source. The use of ground truth points aids in the validation of categorized photos. The overall accuracy attained for 2015 was 96.9%, as shown in Figure 7.3.

By offering high-resolution, current reference data, Google Earth Pro and ERDAS Imagine enhance the accuracy evaluation procedure. This guarantees improved decision-making in RS applications and increases the accuracy of land cover classification. The sum of correctly categorized pixels to all reference pixels is the overall accuracy. The proportion of correctly identified pixels to all actual pixels for that class is known as the producer accuracy (Chauhan, et al., 2020). The ratio of correctly classified pixels to all classified pixels for that class determines the user's accuracy. The accuracy of classification to reference data is 97.5%, as shown in Figure 7.4.

The various classifications of land cover were determined for the study area after the launch of Google Earth Pro. The Place Mark tool was used to digitize sample sites or ground control points (GCPs). Make a KML/KMZ file

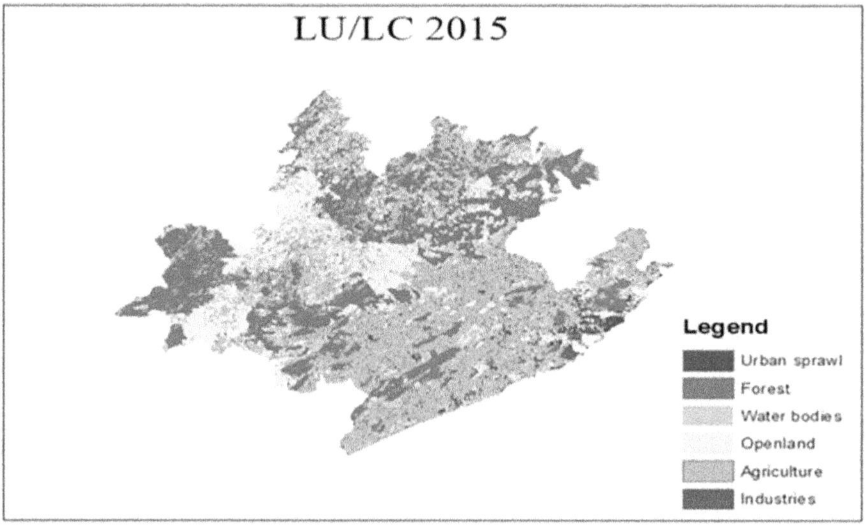

FIGURE 7.2
Land use/land cover of 2015.

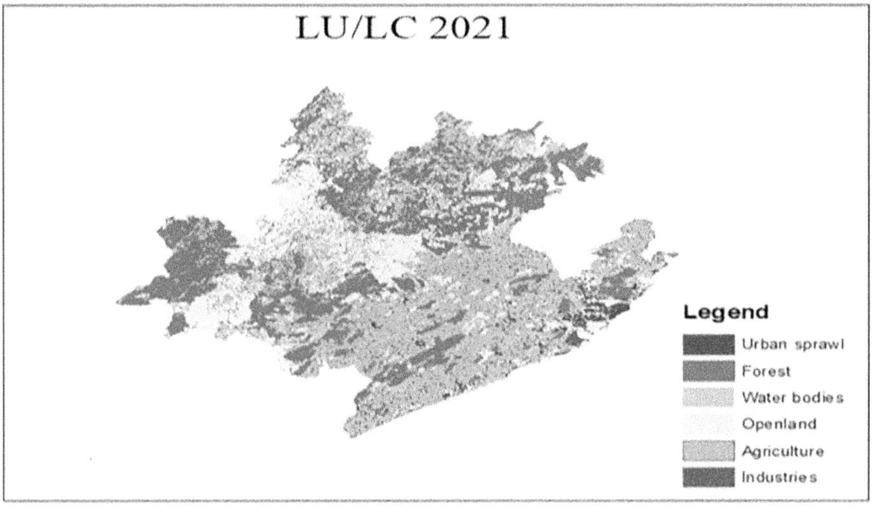

FIGURE 7.3
Land use/land cover of 2021.

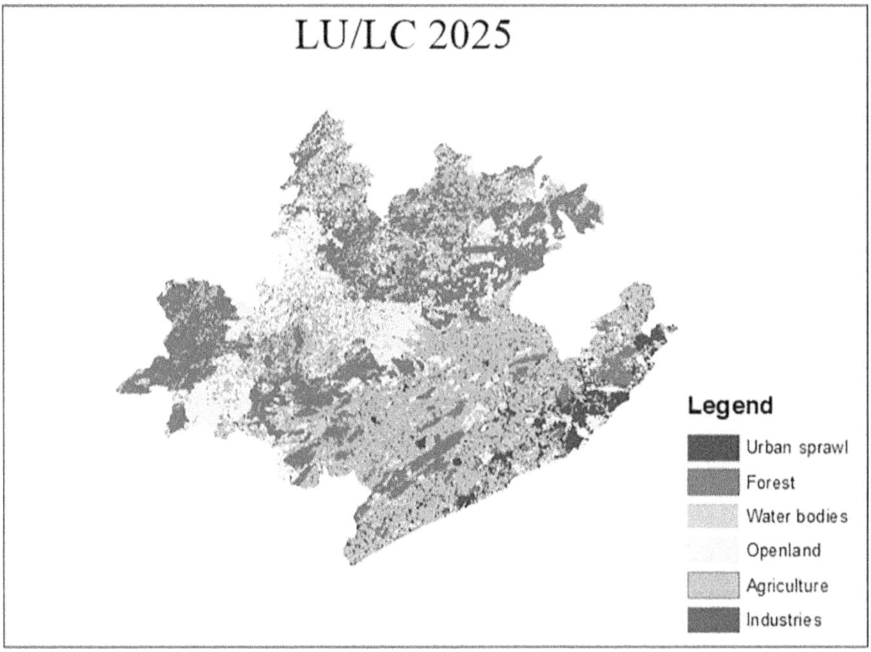

FIGURE 7.4
Land use/land cover of 2025.

TABLE 7.2

Land use/land cover consequences in the study area

LULC Category	Year 2015 Area %	Year 2021 Area %	Year 2025 Area %
Urban sprawl	3.706341	3.922659	7.391865
Forest	27.86557	27.8604	27.63783
Water	3.587594	3.573293	3.395241
Open land	21.7312	21.70766	21.08074
Agriculture	42.13384	41.96053	39.5255
Industries	0.975465	0.975452	0.968815
Total	100	100	100

out of these reference points. GIS software was used to transform the KML/KMZ file into a shapefile. Within ERDAS, Imagine the Google Earth Pro and load it. Reference points from the categorized data. Verify that the coordinate system corresponds to the image that is categorized. Accuracy evaluation is performed for reference points and categorized images in Erdas. A confusion matrix that contrasts the values of the classified and reference pixels was created. The 2025 accuracy rate is 98.4%. as shown in Table 7.2.

7.5 Conclusions

An accurate LC map simulates a crucial aspect of communicating contemporary environmental issues and agronomic requirements. To measure and identify the LULC impacts in the Visakhapatnam region during a 10-year period from 2015 to 2025, RS and GIS were combined in this study. In natural asset management, RS, GIS, and satellite-derived imagery are powerful technologies that are frequently used. The authors of earlier research achieved a maximum accuracy of 95% and a kappa coefficient of 0.94 by using parallelepiped, minimum distance, and maximum likelihood classifier methods. This study uses a classification method that may involve vector digitization and uses pixel classification for satellite images from three different years 2015, 2021, and 2025 to detect changes in land cover. This method yields more accurate results than maximum likelihood classification does. The method used in this study is simple and reasonably priced. Multitemporal satellite imagery was used to determine the degree of land-use alterations in the research area, and a confusion matrix was used to calculate the class accuracy. During this 15-year period, areas beneath vegetation and built-up land increased significantly, whereas areas beneath agricultural land and water bodies decreased. The LULC results in the side survey area account for the decline in agricultural areas and the increase in urban areas. Through in situ measurements obtained from the same geographical location, the RS and GIS were verified. In the research region, the LULC changes did not have a significant environmental impact. Nonetheless, the environmental viability of LULC changes should be closely observed. Additionally, the paintings are most likely dispersed by taking advantage of multiple machines and deep learning algorithms that would enhance the algorithm's performance.

Acknowledgment

The authors are sincerely thankful to our prestigious institute, G. Pulla Reddy Engineering College, and the coauthors who are involved in this study for their direct and indirect contributions.

References

Ademiluyi, I. A, Okude, A. S., & Akanni, C. O. (2008). An appraisal of landuse and landcover mapping in Nigeria. *African Journal of Agricultural Research*, 3(9), 581–586.

Chauhan, N., Kumar, V., & Paliwal, R. (2020). Quantifying the impacts of decadal landuse change on the water balance components using soil and water assessment tool in Ghaggar river basin. *SN Applied Sciences*, 2(11), 1777. https://doi.org/10.1007/s42452-020-03.

Kanmani, K., Padmanabhan, V., & Pari, P. (2023). Accuracy assessment of different classifiers for sustainable development in landuse and landcover mapping using sentinel SAR and landsat-8 data. *EAI Endorsed Transactions on Energy Web*, 10. https://doi.org/10.4108/ew.4141.

Kathwas, A. K., & Patel, N. (2019). Effects of landuse and landcover dynamics on the migration and demographic pattern of tribal population using geospatial techniques. *Theoretical and Empirical Researches in Urban Management*, 14(3), 77–97.

Kotoky, P., Dutta, M. K., & Borah, G. C. (2012). Changes in landuse and landcover along the Dhansiri River channel, Assam – A remote sensing and GIS approach. *Journal of the Geological Society of India*, 79(1), 61–68. https://doi.org/10.1007/s12594-012-0002-6.

Marić, I., Panđa, L., Faričić, J., Šiljeg, A., Domazetović, F., & Marelić, T. (2022). Long-term assessment of spatio-temporal landuse/landcover changes (LUCCs) of Ošljak Island (Croatia) using multi-temporal data invasion of Aleppo Pine. *Land*, 11(5), 620. https://doi.org/10.3390/land11050620.

Paul, A., & Bhattacharji, M. (2023). Prediction of landuse/landcover using CA-ANN approach and its association with river-bank erosion on a stretch of Bhagirathi River of Lower Ganga Plain. *Geo Journal*, 88(3), 3326–3346. https://doi.org/10.1007/s10708-022-10814-1.

Rastogi, U., Kumar, S., Maurya, S. P., & Yadav, A. K. 2025. Air pollution forecasting using genetic algorithm and LSTM deep learning technique. *Aerosol Science and Engineering*. 10.1007/s41810-025-00343-y.

Sajib, A. M., & Moniruzzaman, M. (2022). Driving forces of landuse and landcover changes in the north-eastern part of Dhaka conurbation. *The Dhaka University Journal of Earth and Environmental Sciences*, 10(2). https://doi.org/10.3329/dujees.v10i2.57515.

8

Hybrid Deep Learning Framework Sustainable Environmental System: Solar Irradiance Prediction, Case Study

Madderla Chiranjeevi, Tukaram Moger, and Debashisha Jena

8.1 Introduction

The growing global interest in sustainable energy solutions has led to notable progress in the advancement of solar energy technologies. Solar irradiance forecasting is crucial for effectively integrating photovoltaic power into the existing energy structure. Accurate solar irradiance forecasts support efficient energy management, reduce dependence on conventional power sources, and foster a sustainable future. The inherently variable and complex nature of solar irradiance presents significant challenges to accurate forecasting. As solar power output fluctuates with changes in irradiance, these inconsistencies can affect the stability of electrical grids. Reliable forecasting, therefore, plays a pivotal role in helping grid operators anticipate such fluctuations and effectively manage the integration of solar energy, ultimately contributing to a more stable and dependable power supply.

Various forecasting strategies have been developed over time, ranging from conventional statistical models to the latest machine learning (ML) and hybrid techniques. Forecasting global horizontal irradiance (GHI) is typically categorized according to the type of input data utilized and its prediction time horizon. Traditional approaches, e.g., autoregressive integrated moving average (ARIMA) and persistence models, are popularly used; however, they tend to struggle with capturing the nonlinearity and intricate dynamics typical of solar irradiance data (Bacher et al., 2009; Yang et al., 2012 Yadav, 2025).

Recently, ML, particularly neural network-based models, has shown great promise in enhancing prediction accuracy. Long short-term memory (LSTM) networks, a type of recurrent neural network, are specifically suitable for modeling temporal dependencies. Bidirectional LSTM (BiLSTM) architectures further improve upon this by capturing both past and future contexts, yielding superior performance compared with unidirectional variants (Chiranjeevi

DOI: 10.1201/9781042013234-10

et al., 2023; Rastogi et al., 2025). Nevertheless, these models often fall short in regard to using spatial characteristics embedded within the data.

To address the limitations of standalone models, ensemble and decomposition-based techniques have been introduced. Methods such as empirical mode decomposition (EMD), ensemble empirical mode decomposition (EEMD), and their refined version, complete ensemble empirical mode decomposition with adaptive noise (CEEMDAN), have proven effective in breaking down complex, non-stationary time series into intrinsic mode functions (IMFs), which makes the data more tractable for forecasting models (Huang et al., 1998; Flandrin et al., 2004; Maurya et al., 2022; Karijadi et al., 2022). The CEEMDAN-CNN-LSTM framework, which is designed for hourly GHI prediction, has demonstrated improved performance across various geographical locations (Gao et al., 2020).

Recent advancements have also leveraged hybrid deep learning (DL) architectures. Tong et al. (2022) suggested an innovative encoder-decoder-based hybrid architecture that integrates CEEMDAN with a dynamic error compensation mechanism, demonstrating its efficacy in managing the inherent complexity of solar irradiance data. Similarly, Abumohsen et al. (2024) hybridized convolutional neural network (CNN), LSTM, and random forest (RF) algorithms to forecast solar power, whereas Bai et al. (2025) implemented wavelet packet decomposition alongside LSTM for day-ahead predictions.

In another line of work, Monjoly et al. (2017) explored AR-NN models enhanced by decomposition techniques for GHI forecasting and compared their results with those of traditional approaches to validate the superiority of hybrid models. Li et al. (2025) proposed a feature transformation strategy aimed at minimizing the effects of daily geometric variations. Furthermore, Nahid et al. (2023) introduced a three-step-ahead forecasting model that combines EMD with a CNN and LSTM. Liu et al. (2024) extended this by developing a multimodal decomposition strategy using both Variational Mode Decomposition (VMD)and Complete Ensemble Empirical Mode Decomposition with Adaptive Noise (CEEMDAN) with a Convolutional Neural Network (CNN) - Bidirectional Long Short-Term Memory (BiLSTM) network, achieving high-accuracy photovoltaic forecasts. Despite these advancements, one trade-off remains: the more IMFs used as model inputs, the greater the computational complexity.

Handling missing values is another ongoing challenge. Zhang et al. (2024) addressed this issue via a decomposition-based hybrid transformer model, which yielded high-precision forecasts even when the dataset was incomplete. In addition to solar applications, entropy-based metrics have gained traction for quantifying the unpredictability of time series. Syama, et al. (2024) demonstrated the utility of permutation entropy (PE) in analyzing solar irradiance complexity. Liang et al. (2020) applied PE-enhanced EEMD with regularized extreme learning machines for wind speed prediction, whereas Liu et al. (2024) incorporated VMD and sample entropy to reconstruct wind speed signals before applying BiLSTM models. In a different domain, Zhang et al. (2015) combined PE, EEMD, and SVM for diagnosing bearing faults. Zhang et al. (2024) also applied a VMD and CNN-BiLSTM combination for solar power forecasting, achieving enhanced accuracy at the expense of increased model complexity. Nhat et al. (2024) proposed using EEMD with gated recurrent units (GRUs), further diversifying the hybrid model landscape.

Despite these developments, traditional and ML models often face challenges such as multiscale variability, risk of overfitting, and interpretability issues. Decomposition-based hybrid DL models—especially those integrating CEEMDAN—remain underexplored and present promising opportunities. By isolating noise and extracting meaningful trends through IMFs, these models can incorporate valuable data insights and offer greater scalability, ultimately leading to more robust and reliable solar irradiance forecasting systems.

8.1.1 Contributions of this Study

In contrast to prior work in the domain of irradiance forecasting, this study introduces a set of methodological innovations aimed at addressing key gaps left by existing models and techniques. The key contributions of this chapter are as follows:

1. **Robust Outlier Detection and Intelligent Imputation:**

   This study adopts a two-pronged approach to handle anomalies in the dataset—an area often overlooked or addressed via simplistic methods in past research. First, the isolation forest (IF) algorithm is employed to detect outliers. Compared with conventional statistical techniques, the IF offers enhanced robustness and computational efficiency, making it well suited for high-dimensional, complex solar irradiance data. Following detection, *K*-nearest neighbors (KNNs) are used for data imputation. Unlike traditional mean or interpolation methods, KNN preserves the local structure of the data, thereby maintaining its integrity. The combined application of IF and KNN within the context of solar irradiance forecasting is relatively novel and has yet to be comprehensively explored in the literature.
2. **Application of an Advanced Decomposition Strategy:**

   To handle the persistent issue of mode mixing, a common limitation in earlier decomposition methods, this chapter employs the CEEMDAN method. Although techniques such as EMD and EEMD have been used previously, CEEMDAN offers a more refined decomposition by integrating adaptive noise to separate the signal into IMFs more reliably. This leads to a cleaner and more stable representation of solar irradiance variations, which directly contributes to the accuracy of subsequent forecasting stages.

3. **Innovative feature extraction using PE:**

   This work introduces a new angle for feature extraction by using the PE to rate the complexity of each IMF. What distinguishes this approach is the merging of IMFs into subsets on the basis of their similarity in PE values. Rather than processing each IMF individually—an approach that often introduces redundancy and increases the computational burden—this method simplifies the model architecture while preserving essential information. The reduction in model complexity, without sacrificing predictive power, is a key innovation of this work.

4. **Development of an Integrated Forecasting Framework:**

   The proposed framework has multiple strengths: CEEMDAN for signal decomposition, PE for complexity-based feature extraction, CNN for capturing local spatial patterns, and BiLSTM for modeling temporal dependencies. What sets this framework apart is its approach to forecasting not individual IMFs, but aggregated subsequences formed via entropy-based merging. This holistic integration significantly enhances the prediction accuracy and represents an innovative perspective for handling decomposed series in the case of irradiance prediction.

The chapter is organized as follows: **Section 8.2** outlines the theoretical foundation that forms the basis of the proposed methodology. **Section 8.3** details the experimental setup, including comprehensive information on the datasets utilized, the preprocessing steps undertaken, and the selection of hyperparameters. **Section 8.4** presents the results, offering a thorough analysis and interpretation of the performance of the proposed framework. Finally, **Section 8.5** concludes the chapter with key insights, summarizing the main contributions and suggesting potential future directions.

8.2 Theoretical Discussions

This section elevates the foundational concepts behind the proposed methodology, focusing first on the decomposition technique used in the framework.

8.2.1 Complete Ensemble Empirical Mode Decomposition with Adaptive Noise

CEEMDAN is a popular signal decomposition technique that builds upon and overcomes the limitations of earlier approaches such as EMD and Ensemble

EMD. One of the key issues in traditional methods is mode aliasing—a condition where oscillatory components with different frequencies are not adequately separated during decomposition (Zhang et al., 2020). CEEMDAN overcomes this limitation by employing a more refined strategy.

In CEEMDAN, white noise is systematically added to the signal at each iteration, enabling a more accurate extraction of IMFs. These IMFs collect localized frequency components within the solar irradiance, whereas the remaining residual signal retains the long-term trend characteristics. As the process progresses, the injected noise tends to cancel out across ensemble averages, effectively minimizing reconstruction errors and enhancing the clarity of the decomposed components. This iterative refinement not only improves decomposition accuracy but also ensures better stability in subsequent forecasting tasks. As a result, the mode mixing problem is greatly diminished; thus, the decomposition process is guaranteed to be accurate. The iterative nature of CEEMDAN, through multiple runs with added white noise, enhances its ability to handle both short-term deviations and long-term trends in solar irradiance. The resulting IMFs provide a multiscale depiction of the original data.

At each level of signal decomposition, the operator $E_n(.)$ represents the nth IMF decomposed via the EMD approach (Nahid et al., 2024). IMF_n refers to the nth IMF derived via CEEMDAN decomposition. The signal $y(t)$ denotes the initial collected signal, whereas $w^j(t)$ represents the additional white noise that has been introduced. The value of j ranges from 1 to M, where M represents the total number of instances in which white noise is included. The mathematics of CEEMDAN are as follows:

1. Eq. 8.1 represents the white noise signal that necessitates decomposition via EMD to obtain the first IMF component and is represented as in Eq. 8.2.

$$y^k(t) = y(t) + \beta_o w^j(t),\ (j = 1,2,\ldots, M) \tag{8.1}$$

$$IMF_1(t) = \frac{1}{M}\sum_{j=1}^{M} IMF_1^j(t) \tag{8.2}$$

2. The residual signal is computed via Eq. 8.3:

$$r_1(t) = y^k(t) - IMF_1(t) \tag{8.3}$$

3. The signal $r_1(t) + \beta_1 E_1(w^j(t))$ is further decomposed into M parts. Then, $IMF_2(t)$ is represented as:

$$IMF_2(t) = \frac{1}{M}\sum_{j=1}^{M} E_1(r_1(t) + \beta_1 E_1(w^j(t)) \tag{8.4}$$

4. For $j=1,2, \dots ., J$, the subsequent phases are decomposed, and steps 1 to 3 are reiterated. The jth residual signal r_j is found as in Eq. 5; then, $IMF_{j+1}(t)$ is given as in Eq. 8.6.

$$r_j(t) = r_{n-1}(t) - IMF_n(t) \tag{8.5}$$

$$IMF_{j+1}(t) = \frac{1}{M}\sum_{j=1}^{M} E_1(r_n(t) + \beta_n E_n(w^j(t))) \tag{8.6}$$

5. The above stages are repeated until the residual cannot be further decomposed, meaning that it has either only one extreme value or a monotonic shape. Finally, the original signal is:

$$y_t = \sum_{j=1}^{M} IMF_j + r(t), \ (j = 1,2,. \dots , M) \tag{8.7}$$

8.2.2 Permutation Entropy

Following the decomposition phase, the PE is adopted to extract informative features from every IMF. The PE measures the complexity and unpredictability of solar irradiance by quantifying the frequency of different arrangements of data points (Bandt et al., 2002). CEEMDAN yields a greater number of IMF components due to solar irradiance unpredictability and non-stationarity. This paper employs the PE algorithm for IMF reorganization to reduce the computational complexity and redundancy of similar features. The PE algorithm is described as follows:

1. Initially, phase space reconstruction (Liang et al., 2020) is conducted on every IMF sequence $\{S(j), j=1,2, \dots ., T\}$ derived from CEEMDAN to create a phase space matrix:

$$Y = \{s(i),\ s(i+\tau),\ \dots .,\ s\left(i+(d-1)\tau\right)\} \tag{8.8}$$

where $\tau$and d correspond to the delay time and embedding dimension, respectively, $i=1, 2, \dots, T-(d-1)\tau$.

2. Considering every matrix row as a separate element, a total of K elements are present. The number of column vectors created for *the* M-dimensional vector is $K = M-(d-1)\tau$. For example, consider the ith element in the matrix $s(i+((j_1-1)\tau) = s(i+(j_2-1)\tau)$, in ascending order as per the element value:

$$s\left(j+\left(i_1-1\right)\tau\right) \le s(j+\left(i_2-1\right)\tau$$
$$\le \dots \le s(j+(i_d-1)\tau \tag{8.9}$$

where $i_1, i_2, \dots\dots, i_d$ denotes the serial number of the elements. It is $s\left(j+\left(i_1-1\right)\tau\right) = s\left(j+\left(i_2-1\right)\tau\right)$ arranged as per the size of i_1 and i_2. Therefore, every row vector in the phase space reconstruction matrix of solar irradiance series S_i can generate a group of sequences $S(l)$.

$$S(l) = \{i_1, i_2, \dots\dots, i_d\} \tag{8.10}$$

3. Compute the relative frequencies or probabilities, $p_1, p_2, \dots, p_T$ in the sequence $S(l)$. It is found via the number of occurrences of the permutation in the sequence divided by whole sequences.
4. Compute the PE of order '*D*' of the solar irradiance series by

$$PE_D = -\sum_{i=1}^{k} p_i.log_2 p_i \tag{8.11}$$

If the PE value of the series is higher, the series is more irregular and random.

The PE of all the IMFs is measured via the above steps. The IMF components are merged to generate a merged subsequence on the basis of the proximity of the PE values.

8.2.3 CNN-BiLSTM Fusion Model

The hybrid CNN–BiLSTM model is designed to exploit the spatial and temporal characteristics present in solar irradiance data. In this framework, the CNN is responsible for capturing spatial patterns and localized features from

PE-enhanced IMFs. By applying convolutional filters across the transformed data, the CNN effectively identifies recurring structures and relationships within the input (Elmousaid et al., 2024). Once the spatial features are extracted, they are passed into the BiLSTM model, which is well suited for learning temporal dependencies. Unlike standard LSTM models, BiLSTM processes information in either direction, enabling it to gather contextual information across time more comprehensively.

By aggregating the capabilities of the CNN and BiLSTM models, the proposed hybrid framework effectively utilizes a CNN to extract spatial characteristics and BiLSTM to capture temporal dependencies (Chiranjeevi et al., 2024). This integration improves the system's ability to manage both immediate variations and extended patterns in solar irradiance data, making it a strong candidate for handling complex forecasting tasks in practical-environment.

The CNN is the first part of the hybrid model. The CNN comprises an input layer that receives inputs, an output layer that extracts features for the BiLSTMs, and many hidden layers. The hidden layer comprises convolution, pooling, and flattening layers. The convolution layer performs the convolution operation on the input solar irradiance and passes the output to the subsequent layer. The convolution procedure emulates the response of individual neurons to visual stimuli. Every convolutional neuron exclusively analyses the merged IMF inside its receptive area.

If $s_i^o = \{s_1, s_2, \ldots, s_n\}$ is the merged IMF sequence used as the input vector, the output vector y_{ij}^l after the convolutional and hidden layers l is:

$$y_{ij}^l = \sigma\left(b_j^l + \sum_{m=1}^{M} w_{m,j}^l \, s_{i+m-1,j}^o \right) \tag{8.12}$$

where σ denotes the activation function, b_j^l represents the bias for the jth feature map, m is the filter index, and w is the kernel weight.

The BiLSTM is a submodel that follows the CNN in the CNN-BiLSTM architecture to capture the temporal features of the data (Wang et al., 2024). The BiLSTM model employs two LSTM units to understand the information contained in each sequence. It considers the past and future series. As illustrated in **Figure 8.1**, one LSTM model processes the sequence in a forward direction, whereas another LSTM model processes it in a backward direction.

The LSTM model is composed of an input gate, an output gate, and a forget gate through which the sequence of data flows in the layers (Chiranjeevi et al., 2023).

$$i_t = \sigma(w_i \left[h_{t-1}, s_t\right] + b_i) \tag{8.13}$$

$$o_t = \sigma(w_o \left[h_{t-1}, s_t\right] + b_o) \tag{8.14}$$

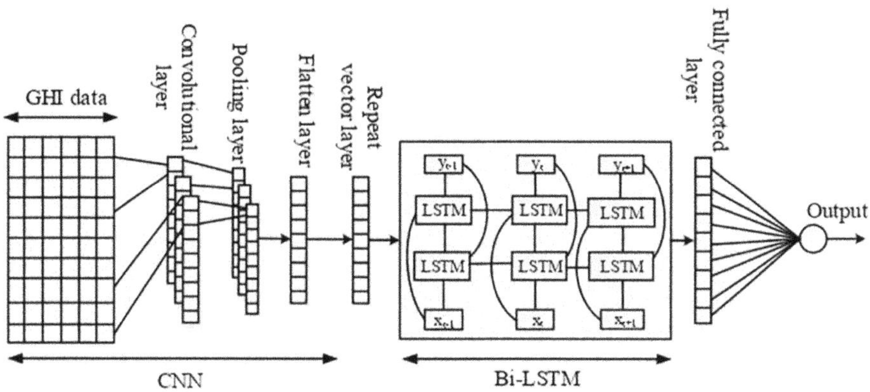

FIGURE 8.1
Fusion of the convolutional neural network-bidirectional long short-term memory architecture.

$$f_t = \sigma\left(w_f\left[h_{t-1}, s_t\right] + b_f\right) \tag{8.15}$$

The candidate state, cell state, and output state are determined via Eqs. (8.16), (8.17), and (8.18) after the above values are obtained. At time t, c_t represents the cell memory, and $\widetilde{c_t}$ represents the candidate cell.

$$\widetilde{c_t} = tanh(w_c\left[h_{t-1}, s_t\right] + b_c) \tag{8.16}$$

$$c_t = f_t * c_{t-1} + i_t * \overline{c_t}) \tag{8.17}$$

$$h_t = o_t * \tanh(c_t) \tag{8.18}$$

8.2.4 Proposed Method

After all the above stages are hybridized and implemented, the predicted aggregated sequence is generated, which is compared with the actual test solar irradiance data to confirm the model's effectiveness. The flow chart of the proposed methodology is depicted in **Figure 8.2**, through which the step-by-step implementation of the model can be understood.

The proposed approach for solar irradiance prediction involves the following steps:

a. **Data collection and preprocessing**

   Solar irradiance data are initially gathered from reliable sources. Preprocessing is performed to manage missing values and detect outliers. Techniques such as IF for outlier detection and KNN for imputation are applied to assure the dataset's quality and consistency.

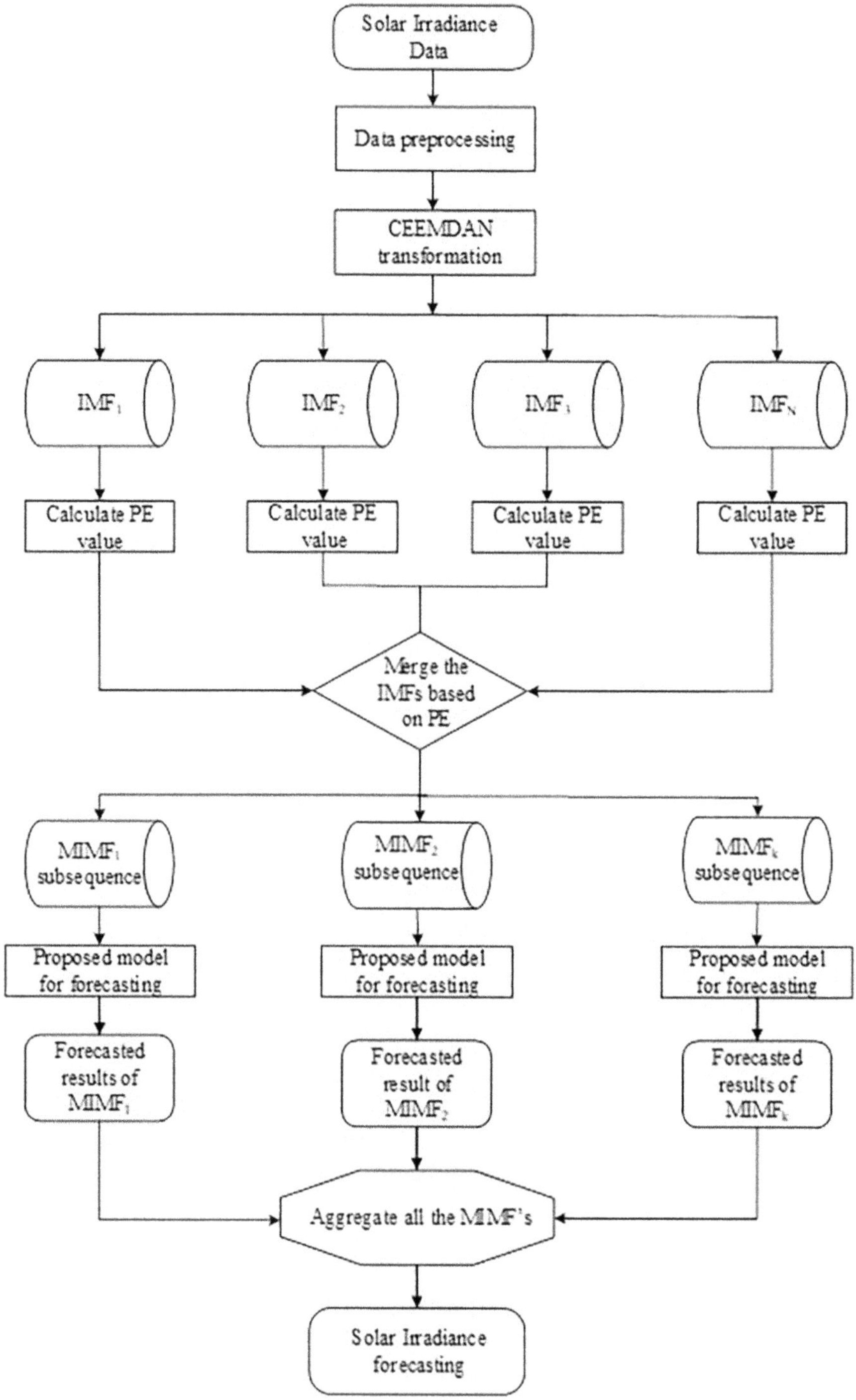

FIGURE 8.2
Process of implementation of the proposed framework.

b. **CEEMDAN decomposition**

The cleaned data are decomposed via the CEEMDAN technique. This process breaks down the series into a group of IMFs, each representing oscillatory components at different frequency levels.

c. **PE computation**

PE is calculated for each IMF to assess the complexity and dynamical variability of the signals. PE serves as a basis for understanding the informational content and behavior of each mode.

d. **IMF Merging Based on PE Similarity**

IMFs exhibiting similar PE values are grouped together into merged intrinsic mode function sequences (mIMFs). This step reduces redundancy, improves interpretability, and ensures that structurally similar modes are treated jointly in the forecasting process.

e. **Subsequent Forecasting via Hybrid Deep Learning**

Each mIMF is treated as a distinct input sequence. The proposed hybrid model is applied to forecast these subsequences. The CNN module captures spatial relationships, whereas the BiLSTM network models temporal dynamics, enabling the framework to learn complex nonlinear patterns.

f. **Aggregation of Forecasted Sequences**

Forecast outputs from all mIMFs are aggregated to form the final prediction of solar irradiance. This step reconstructs the forecasted signal by summing the individual components.

g. **Performance evaluation**

The accuracy and reliability of the forecast are assessed via error metrics obtained from actual and predicted values. Comparative analysis is conducted against baseline models to show the effectiveness of the proposed framework.

8.3 Experimental Considerations

8.3.1 Data Collection

The research is carried out via datasets obtained from the Hawai'i Space Exploration Analog and Simulation Research Station (HI-SEAS) and the National Institute of Technology Karnataka (NITK) campus, Surathkal, India. An image of the data collection site situated at NITK is presented in **Figure 8.3**, which is positioned at the top of the Electrical Department of NITK. The statistical details of the datasets are given in **Table 8.1**, through which the statistical

TABLE 8.1

Statistical details of the Solar irradiance

Data Center	Min. Value	Max. Value	Mean Value	Std. Dev
HI-SEAS	0.00	1601.26	207.124	315.916
NITK	0.00	1340.60	198.863	287.093

Note: The units are in W/m^2

FIGURE 8.3
National Institute of Technology Karnataka data collection center located on an EEE department rooftop.

parameters can be observed. The HI-SEAS dataset comprises 32,686 samples taken at 5-minute intervals over a 4-month period, beginning on September 29, 2016, at 12:00 and ending on December 1, 2016, at 00:00. The NITK dataset comprises 39,654 samples collected from September 1, 2023, from 00:00 to January 16, 2024, at 16:30. Overall, 80% of the data are allocated for training purposes, with the remaining 20% meant for testing. All the simulations are conducted on a desktop system with an Intel Core i7-10700 CPU @ 2.90 GHz, 16.00 GB of RAM, and the Python 3.7 programming language.

8.3.2 Data Preprocessing

The GHI values may consist of missing and extreme values primarily arising from intermittent inverter issues, system disconnection for cable repairs, replacement of damaged modules and data outages. The raw information obtained from the PV centers needs to undergo imputation for missing values and outlier checking. The raw data consists of several negative values in the morning hours and night hours, which are replaced with zeros. The dataset consists of several outliers that are observed via IF (Liu et al., 2008).

The IF algorithm is an unsupervised technique employed in the collective-based model to identify anomalies by calculating the isolation score for each data point. The IF algorithm is based on the same fundamental principles as decision tree ensembles, particularly those used in the RF method. However, instead of focusing on classification or regression, the IF is particularly useful for anomaly detection. It functions by recursively splitting the time series into random subsets using randomly selected features and splitting values. These split values are chosen within the feature's observed range, i.e., between its minimum and maximum values. Data points that are isolated with fewer splits are more likely to be outliers, as they deviate from the majority distribution. Once the outliers are identified, they are handled via the K-nearest neighbor (KNN) imputation method. This approach estimates the missing or anomalous values by referencing the most similar data in the feature space, thereby preserving the statistical structure of the dataset and enhancing the effectiveness of the forecasting model (Zou et al., 2023).

Normalizing the solar irradiance data is a critical preprocessing step that transforms the input features into a standardized scale. This process ensures that all variables contribute equally during model training, reduces computational complexity, and often leads to improved predictive accuracy. The normalization procedure applied in this study is expressed in Eq. 8.19.

$$x' = \frac{(x - x_{min})}{(x_{max} - x_{min})} \tag{8.19}$$

where x_{min} and x_{max} are the lowest and highest values of the GHI.

8.3.3 Evaluation Metrics Used

The proposed model is formulated and tested with real-world solar irradiance data. During the training phase, historical data are used to optimize the model parameters, ensuring that the architecture learns the underlying temporal and spatial patterns. In the evaluation phase, the model's forecasting ability is assessed on a separate test set to gauge its generalization performance. To quantify the accuracy and effectiveness of forecasts, several widely accepted performance metrics are employed, which include the mean absolute error (MAE), root mean square error (RMSE), and coefficient of determination (R^2 score) (Chiranjeevi et al., 2023). Given the nonstationary and nonlinear nature of solar irradiance, a specific metric may not provide a comprehensive evaluation. Therefore, multiple error metrics are used in parallel to perform a well-rounded performance assessment. A concise summary of the evaluation criteria applied in this chapter is presented in **Table 8.2**.

8.3.4 Hyperparameters of the Proposed Framework

The hyperparameters of the proposed framework are known through a combination of grid search and cross-validation techniques. This integrated approach facilitates the selection of optimal parameter values by systematically evaluating performance across multiple configurations, thereby enhancing both the predictive accuracy and the model's generalization ability. This strategy could mitigate concerns regarding the fixed hyperparameter method while attaining robust performance across multiple settings. The optimum values of the proposed framework obtained from the grid search and cross validation are given in **Table 8.3**.

TABLE 8.2

Evaluation Metrices

Metric	Representation	Definition
MAE	$\frac{1}{N}\sum_{i=1}^{N}\left\|y_i-p_i\right\|$	Calculates the average absolute error of N forecasting results.
MSE	$\frac{\sum_{i=1}^{N}(y_i-p_i)^2}{N}$	Calculates average of error squares.
RMSE	$\sqrt{\frac{\sum_{i=1}^{N}(y_i-p_i)^2}{N}}$	Determines the square root of MSE.
R^2 score	$1-\frac{\sum_{i=1}^{N}(y_i-p_i)^2}{\sum_{i=1}^{N}(y_i-\bar{y})^2}$	Represents the proportion of variance in output variable.

TABLE 8.3
Optimal Hyperparameters of the Proposed Model

Component	Hyperparameter	Optimal Value
CEEMDAN	Ensemble size	100
	Noise amplitude	0.2 (s.d)
	Max iterations	50
PE	Embedding dimensions	3
	Time delay	1
CNN	Number of layers	2
	Number of filters	128,64
	Kernel size	2
	Pooling size	2x2
	Activation function	ReLU
BiLSTM	Number of units	100,50
	Number of layers	2
	Dropout rate	0.1
Training parameters	Batch size	32
	Learning rate	0.001
	Epochs	100
	Optimizer	Adam

8.4 Results and Discussions

8.4.1 Performance of the Proposed Framework

The efficiency of forecasting can be improved through the implementation of the proposed framework. The CEEMDAN model initially decomposes solar radiation into IMF components, as shown in **Figures 8.4 and 8.5**. These figures clearly show that every IMF component obtained via the decomposition model is different from one another. The decomposition stages are based on randomness in the data and cannot be predefined. The quantity of IMF components may differ depending on the intrinsic attributes of the data. During the decomposition process, the components exhibit increased regularity, leading to more IMFs. The mode mixing problem (i.e., superimposition of one IMF component on another IMF component) in EMD can be eliminated, as evident in **Figures 8.4 and 8.5**. The computational burden increases if every component is predicted individually. The PE is applied to differentiate the difficulty of each IMF, and then the IMFs with equivalent entropy values are reorganized to generate a new mIMF to increase the accuracy and computational speed of solar irradiance forecasting.

The PE values of every decomposition model are represented visually in **Figures 8.6 and 8.7**. The low-order IMFs (high-frequency components) have higher PE values than the higher-order IMFs (low-frequency components),

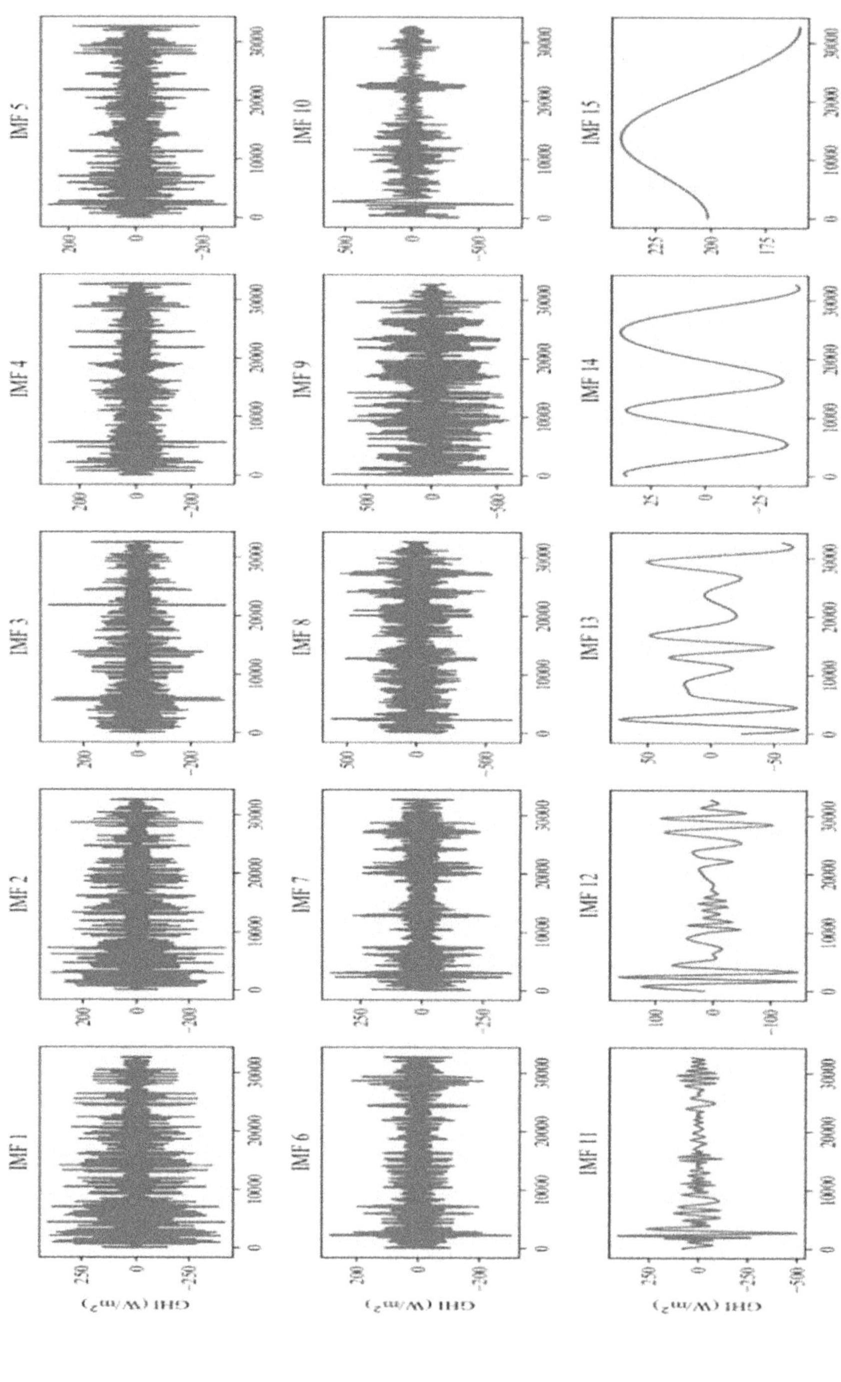

FIGURE 8.4
Intrinsic mode functions obtained via complete ensemble empirical mode decomposition with adaptive noise for Hawai'i Space Exploration Analog and Simulation Research Station data.

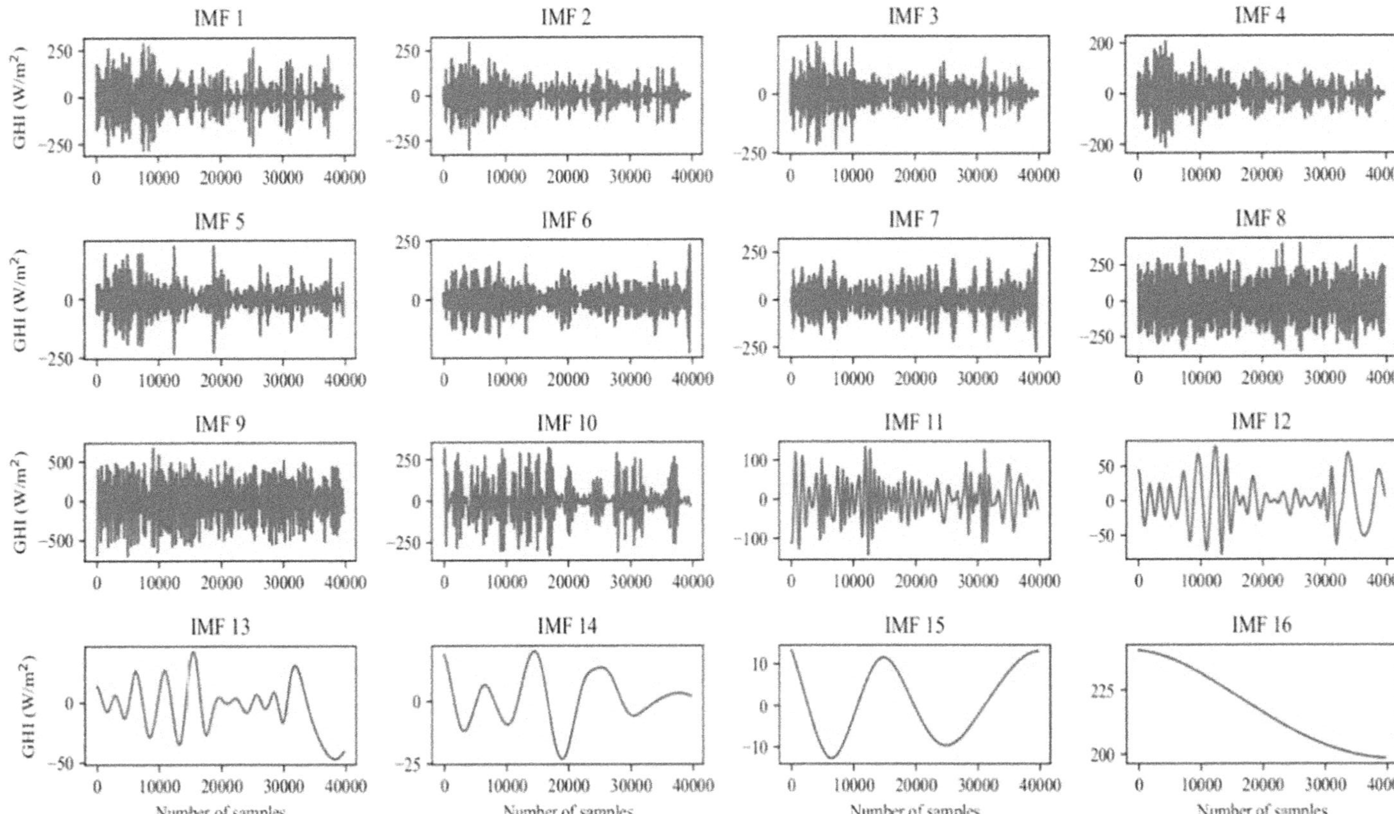

FIGURE 8.5
Intrinsic mode functions obtained via complete ensemble empirical mode decomposition with adaptive noise for National Institute of Technology Karnataka data.

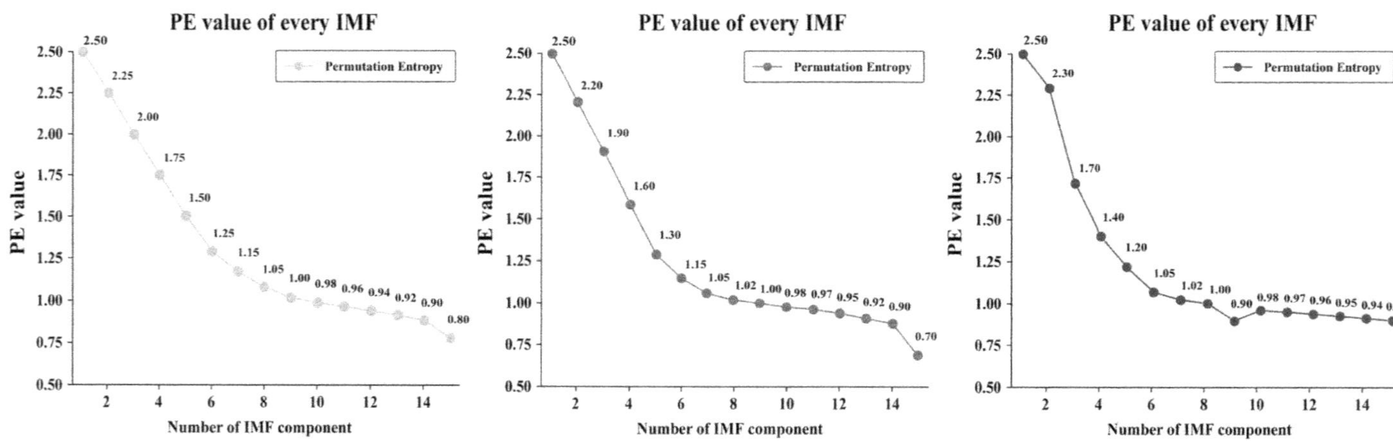

FIGURE 8.6
Permutation entropy values of all the decomposition models for the Hawai'i Space Exploration Analog and Simulation Research Station data. (a) Permutation entropy values of empirical mode decomposition. (b) Permutation entropy values of ensemble empirical mode decomposition. (c) Permutation entropy values of complete ensemble empirical mode decomposition with adaptive noise.

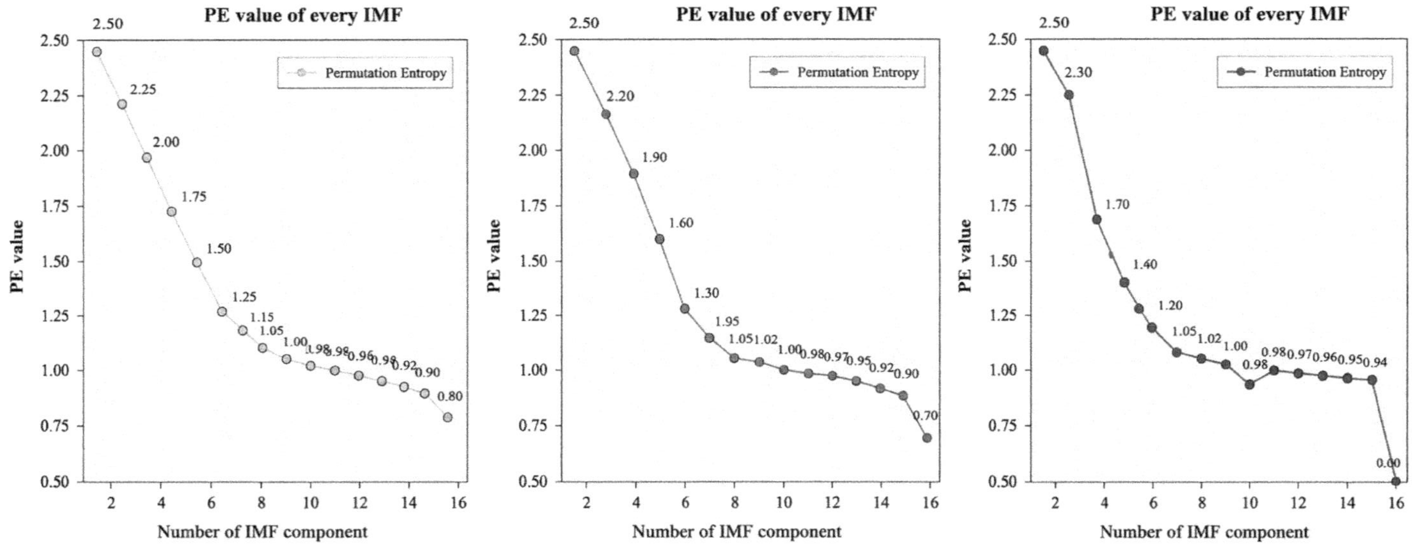

FIGURE 8.7
Permutation entropy values of all the decomposition models for the National Institute of Technology Karnataka data. (a) permutation entropy values of empirical mode decomposition. (b) permutation entropy values of ensemble empirical mode decomposition. (c) permutation entropy values of complete ensemble empirical mode decomposition with adaptive noise.

which have lower PE values. The recombination of the IMFs on the basis of the PE values is provided in **Table 8.4**. If the difference between consecutive IMFs is 0.16, then they are merged into a new IMF. The process is continued in all the decomposition models. The six newly merged IMFs drastically reduce the complexity of the forecasting model without compromising the original characteristics of the solar irradiance data. The higher-frequency IMF components have more uncertainties and are less accurate than the low-frequency IMFs. To overcome these ambiguities and reduce the complexity in the prediction, the IMFs are merged, which results in a higher forecasting accuracy in the proposed model.

The predictions are compared with those of various other hybrid DL architectures; the proposed hybrid framework outperforms other hybrid models. The solar irradiance of the predicted series follows the test data in the proposed model better than that of any other model. The evaluation of conventional models without decomposition is presented in **Table 8.5**. The CNN-BiLSTM performs the best, whereas the decision tree algorithm performs the worst in comparison with all other models w.r.t. error metrics. Forecasting via CEEMDAN-PE-CNN-BiLSTM outperforms all the other models. The evaluation of all the models is improved when it is combined with decomposition techniques, as opposed to a lower performance when decomposition is not implemented, as is evident in **Table 8.5**. When the DL models are combined, the performance is better than that of individual models such as ARIMA, linear regression, RF, decision tree, SVR, KNN, MLPNN, LSTM, XGBoost, BiLSTM, CNN-LSTM, and CNN-BiLSTM.

The introduction of the decomposition technique with PE has a greater impact w.r.t. accuracy. The performance of the proposed model can be inferred from **Figures 8.8 and 8.9**. Compared with the other hybrid architectures, the proposed architecture almost always follows the actual test values. A comparison of various hybrid models implemented with different DL architectures is presented in **Table 8.6 and 8.7**. The CEEMDAN technique in combination with various hybrid DL models performs better than EEMD and EMD in terms of all performance metrics for both one step ahead and multiple steps ahead. As the number of steps in forecasting increases, the accuracy decreases. However, the proposed method performs well in all cases in comparison with the state-of-the-art models.

Compared with models that do not employ decomposition, the forecasting performance is consistently improved when the decomposition is implemented on the input series and similar IMF components are grouped via PE. However, the proposed model outperforms all the other models considered. The MAE, RMSE, and R^2 score of the proposed framework for one step ahead or five minutes ahead are 10.256 (W/m^2), 17.688 (W/m^2), and 0.984, respectively, for the HI-SEAS data. The performance metrics for the NITK data are 5.235 (W/m^2), 13.026 (W/m^2), and 0.997. Similarly, for 10 minutes and 1 hour ahead, the results are 22.265 (W/m^2), 44.497 (W/m^2),

TABLE 8.4

Intrinsic mode function values for the three decomposition models for two datasets

Data	HI-SEAS						NITK					
IMF	**EMD**	**mIMF**	**EEMD**	**mIMF**	**CEEMDAN**	**mIMF**	**EMD**	**mIMF**	**EEMD**	**mIMF**	**CEEMDAN**	**mIMF**
1	2.329	m1	2.577	m1	2.5848	m1	2.571	m1	2.577	m1	2.580	m1
2	2.040	m2	2.408	m1	2.526	m1	2.323	m2	2.404	m1	2.548	m1
3	1.807	m3	2.066	m2	1.975	m2	2.016	m3	2.053	m2	2.069	m2
4	1.626	m4	1.763	m3	1.623	m3	1.769	m4	1.738	m3	1.683	m3
5	1.445	m5	1.511	m4	1.497	m4	1.585	m5	1.469	m4	1.469	m4
6	1.38	m5	1.309	m5	1.331	m5	1.457	m5	1.248	m4	1.345	m5
7	1.289	m5	1.186	m5	1.228	m5	1.343	m5	1.123	m4	1.241	m5
8	1.196	m5	1.125	m5	1.129	m5	1.227	m5	1.072	m4	1.135	m5
9	1.118	m5	1.068	m5	1.072	m5	1.121	m5	1.030	m4	1.041	m5
10	1.066	m5	1.061	m5	1.039	m6	1.072	m5	1.030	m4	1.041	m5
11	1.037	m5	1.023	m5	1.024	m6	1.045	m5	1.014	m4	1.019	m5
12	1.017	m5	1.010	m5	1.011	m6	1.022	m5	1.007	m4	1.010	m5
13	1.007	m5	0.992	m5	1.001	m6	1.012	m5	1.001	m4	1.004	m5
14	0.997	m5	0.964	m5	1.000	m6	1.002	m5	0.929	m4	1.001	m5
15	0.958	m5	0.967	m5	0.994	m6	0.997	m5	0.358	m5	0.995	m5
16	0.631	m6	0.716	m6			0.994	m5	0.106	m6	0.700	m6
17							0.859	m6				

TABLE 8.5

Performance of different statistical and machine learning models without decomposition for one-step ahead

Dataset	HI-SEAS			NITK		
Model	**MAE**	**RMSE**	**R^2**	**MAE**	**RMSE**	**R^2**
ARIMA	19.429	47.581	0.929	16.676	40.744	0.976
Decision Tree	26.110	84.742	0.877	16.041	53.32	0.964
RF	18.792	56.719	0.945	12.414	39.57	0.980
SVR	29.162	68.508	0.919	19.063	45.519	0.974
XGBoost	19.988	58.340	0.941	11.975	37.289	0.976
MLPNN	31.40	82.593	0.930	19.697	54.183	0.964
LSTM	20.35	51.824	0.952	14.834	36.181	0.981
BiLSTM	19.58	48.561	0.955	18.391	38.121	0.981
CNN-LSTM	19.20	45.490	0.956	13.665	36.907	0.982
CNN-BiLSTM	**18.22**	**44.651**	**0.959**	**12.942**	**36.844**	**0.983**

TABLE 8.6

Multi step ahead forecasting evaluation of decomposition models with different hybrid deep learning techniques with Hawai'i Space Exploration Analog and Simulation Research Station dataset

Dataset		EMD-PE			EEMD-PE			CEEMDAN-PE		
Model	**Metrices**	**1 step**	**2 steps**	**12 steps**	**1 step**	**2 steps**	**12 steps**	**1 step**	**2 steps**	**12 steps**
M1	MAE	16.284	46.131	68.722	14.645	45.752	65.596	14.014	43.443	62.744
	RMSE	38.629	95.738	134.120	37.133	93.959	129.845	35.829	90.502	125.594
	R^2	0.951	0.866	0.724	0.965	0.849	0.742	0.971	0.860	0.753
M2	MAE	15.231	41.837	61.894	14.054	41.952	62.953	13.383	46.096	55.772
	RMSE	35.527	63.096	117.414	34.762	61.826	118.484	31.465	58.131	111.343
	R^2	0.963	0.913	0.738	0.966	0.919	0.951	0.976	0.921	0.768
M3	MAE	15.044	31.611	61.514	14.171	29.401	71.341	13.215	27.044	58.746
	RMSE	33.681	58.813	125.777	32.783	57.930	119.080	30.015	52.412	118.258
	R^2	0.965	0.936	0.745	0.969	0.939	0.761	0.979	0.949	0.773
M4	MAE	12.753	24.624	56.908	12.681	23.271	56.210	10.425	21.554	56.700
	RMSE	24.235	56.153	111.757	21.655	55.985	108.853	18.156	51.951	107.604
	R^2	0.972	0.941	0.778	0.976	0.942	0.792	0.980	0.941	0.795
M5	MAE	12.533	23.527	50.083	12.248	22.858	49.411	**10.256**	**22.265**	**48.297**
	RMSE	24.015	49.635	108.364	20.929	54.032	105.399	**17.688**	**44.497**	**102.769**
	R^2	0.975	0.945	0.799	0.978	0.947	0.813	**0.984**	**0.958**	**0.826**

Note 1: *M1 is MLPNN, M2 is LSTM, M3 is BiLSTM, M4 is CNN-LSTM, and M5 is CNN-BiLSTM. All the considered models are in combination with the corresponding type of decomposition.*
Note 2: *MAE and RMSE are in W/m^2 and R^2 score is unitless.*

TABLE 8.7

Multistep-ahead forecasting evaluation of decomposition models with different hybrid deep learning techniques with the National Institute of Technology Karnataka dataset

Dataset		**EMD-PE**			**EEMD-PE**			**CEEMDAN-PE**		
Model	Metrices	1 step	2 steps	12 steps	1 step	2 steps	12 steps	1 step	2 steps	12 steps
M1	MAE	12.301	48.193	69.860	11.062	38.553	62.072	10.199	51.193	59.860
	RMSE	19.514	62.028	98.060	17.510	56.701	82.995	16.953	55.028	81.060
	R^2	0.975	0.934	0.841	0.979	0.941	0.863	0.987	0.946	0.873
M2	MAE	10.401	34.151	67.139	10.334	31.556	56.640	9.981	36.058	52.298
	RMSE	18.602	63.842	87.940	18.061	43.651	83.990	16.399	42.136	79.942
	R^2	0.979	0.941	0.882	0.985	0.948	0.871	0.986	0.952	0.886
M3	MAE	8.718	31.379	64.642	9.695	33.452	54.570	9.594	29.190	50.204
	RMSE	18.466	48.736	86.207	17.653	42.385	81.638	16.055	39.551	77.819
	R^2	0.982	0.968	0.890	0.989	0.963	0.884	0.993	0.980	0.892
M4	MAE	9.311	27.230	46.175	8.950	24.337	54.342	7.141	21.970	49.185
	RMSE	18.293	45.286	84.480	16.895	38.612	84.836	14.488	37.951	76.583
	R^2	0.992	0.974	0.923	0.994	0.976	0.889	0.996	0.982	0.894
M5	MAE	8.330	23.948	43.769	8.089	21.531	49.005	**5.235**	**18.691**	**43.645**
	RMSE	14.489	42.813	83.957	14.096	39.172	79.050	**13.026**	**36.295**	**76.040**
	R^2	0.995	0.977	0.905	0996	0.979	0.919	**0.997**	**0.984**	**0.922**

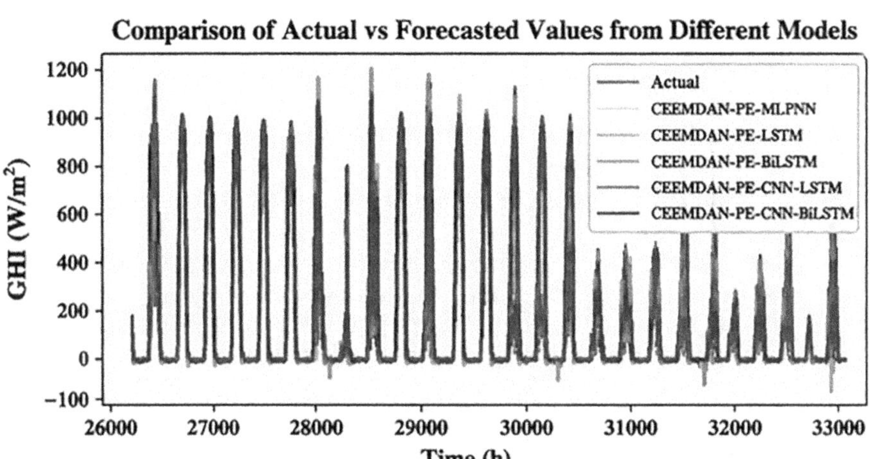

FIGURE 8.8

Comparison of all models using the complete ensemble empirical mode decomposition with adaptive noise for Hawai'i Space Exploration Analog and Simulation Research Station data.

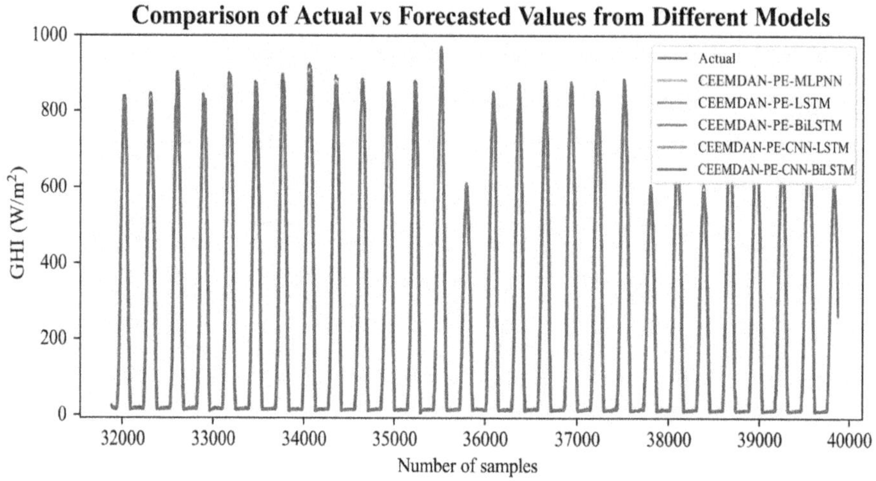

FIGURE 8.9
Comparison of all models using the complete ensemble empirical mode decomposition with adaptive noise for National Institute of Technology Karnataka data.

0.958 and 48.297 (W/m^2), 104.769 (W/m^2), and 0.826 for the HI-SEAS data, whereas for the NITK data, the results are 18.691 (W/m^2), 36.295 (W/m^2), 0.984 and 43.645 (W/m^2), 76.040 (W/m^2), and 0.922, respectively.

8.4.2 Error Analysis of the Proposed Framework

The chapter presents a comprehensive analysis of solar radiation forecasting results across various forecast horizons, spanning from five minutes ahead (one step) to one hour ahead (12 steps). The study begins by comparing the accuracy and robustness of different forecasting models and assessing their performance at each forecast horizon. **Figures 8.10 and 8.11** reveal notable discrepancies in model performance as the forecast horizon extends. These figures depict comparisons across the RMSE, MAE, and R^2 scores for one-step, two-step, and 12-step ahead forecasting. Notably, it becomes evident that forecasting across a horizon of one step is superior to that of two-steps and 12 steps ahead. Moreover, the study explores potential overfitting in the models, particularly concerning long-term forecasting, and deliberates on the implications of uncertainty associated with extended-term solar irradiance forecasts.

Figure 8.12 clearly shows that the proposed forecast model effectively forecasts solar irradiance and monitors daily fluctuations in the GHI. Compared with other forecasting frameworks, the proposed model has a distinct advantage. The MLPNN model exhibits significant variation in deviation, with a notable forecast bias ranging from -7.671 to 889.091 W/m^2. Compared with the proposed models, the LSTM and BiLSTM models

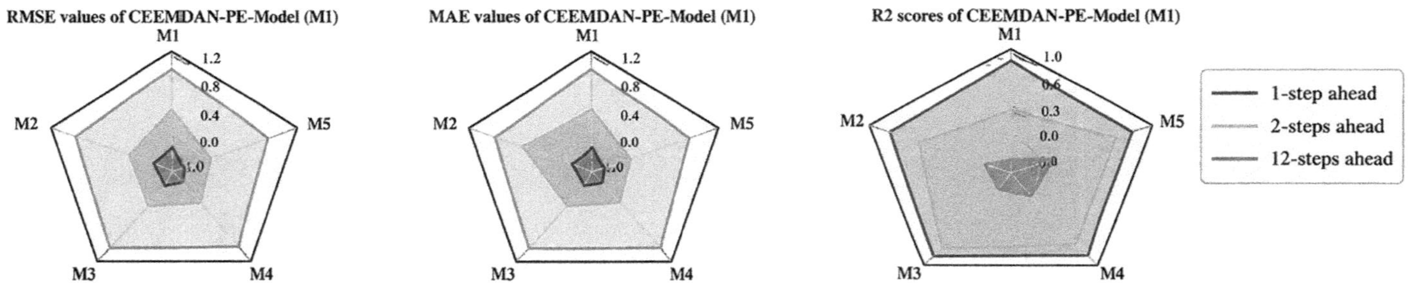

FIGURE 8.10
Error analysis of the proposed model for multistep-ahead forecasting of Hawai'i Space Exploration Analog and Simulation Research Station data.

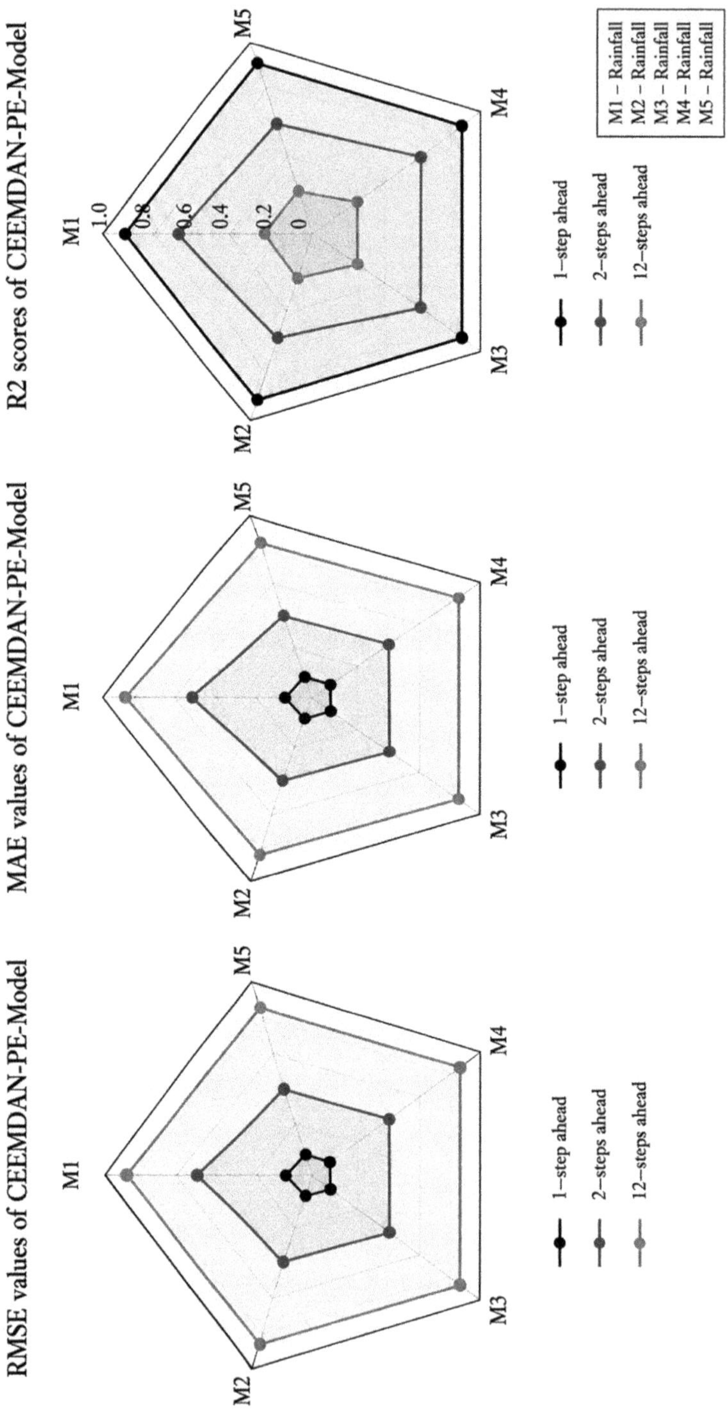

FIGURE 8.11
Error analysis of the proposed model for multistep-ahead forecasting of National Institute of Technology Karnataka data.

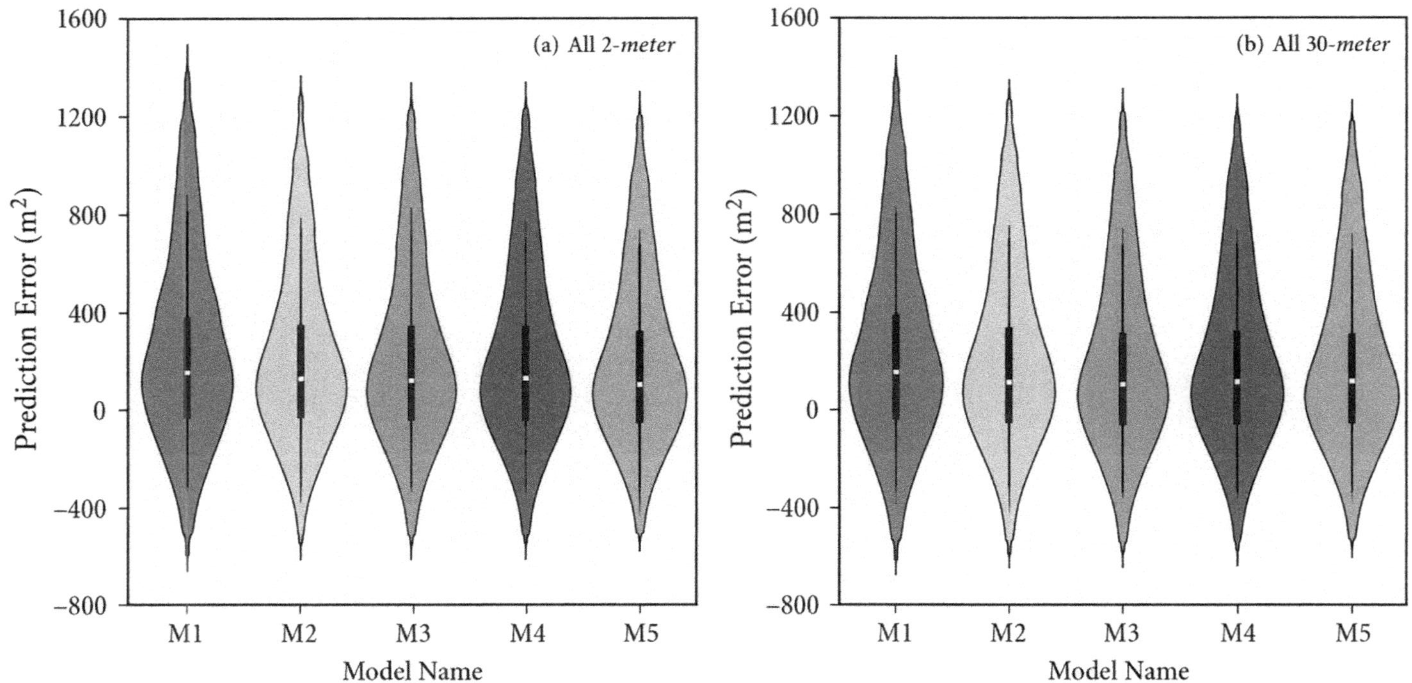

FIGURE 8.12
Standard deviations of the predictions of all the models via the complete ensemble empirical mode decomposition with adaptive noise-permutation entropy model (a) for the Hawai'i Space Exploration Analog and Simulation Research Station data. (b) National Institute of Technology Karnataka data.

yield more substantial forecast deviations in both datasets considered. The CEEMDAN-PE-BiLSTM has the best accuracy and the least forecast deviation, which is [-2.258, 805.697 W/m^2]. The CEEMDAN-PE-LSTM model has a comparable forecast accuracy to that of the proposed model, with a forecast deviation ranging from -6.788 to 820.907 W/m^2. Additionally, the scatter plots of the CEEMDAN method with other hybrid DL models are represented in **Figures 8.13 and 8.14.** From these figures, it is evident that the proposed framework has low deviations of actual vs. predicted GHI values compared to other forecasting models. The findings underscore that hybrid models incorporating decomposition techniques consistently outperform other hybrid approaches in terms of predictive accuracy.

8.5 Conclusion

This study outlines an advanced methodology for solar irradiance forecasting by hybridizing CEEMDAN, PE, and a hybrid CNN-BiLSTM model. By combining decomposition, entropy-guided feature extraction, and DL-based spatiotemporal modeling, the proposed approach significantly improves the forecasting precision, particularly for multistep ahead predictions. Comprehensive experiments conducted on real-world data validate the model's efficiency. The evaluation includes a comparison of the model's predictive ability with and without decomposition, as well as with alternative decomposition strategies such as EMD and EEMD. On the NITK dataset, the proposed framework achieves an MAE of 5.235 W/m^2, an RMSE of 13.026 W/m^2, and an R^2 score of 0.997 for one-step ahead forecasting—highlighting its superior accuracy and robustness relative to baseline models.

This work contributes meaningfully to the domain of renewable energy forecasting. The improved forecasting capability not only enhances operational decision-making for grid operators and energy planners but also supports more reliable integration of solar energy into power systems. The methodology's adaptability and performance make it a promising candidate for real-time deployment in smart grid environments. In the future, several avenues for improvement are evident. These include exploring more sophisticated strategies for IMF grouping and selection, adapting the framework to a wider range of meteorological conditions, and extending its application to forecast other renewable sources. By addressing these challenges, the proposed architecture can progress into a comprehensive solution for data-driven renewable energy forecasting.

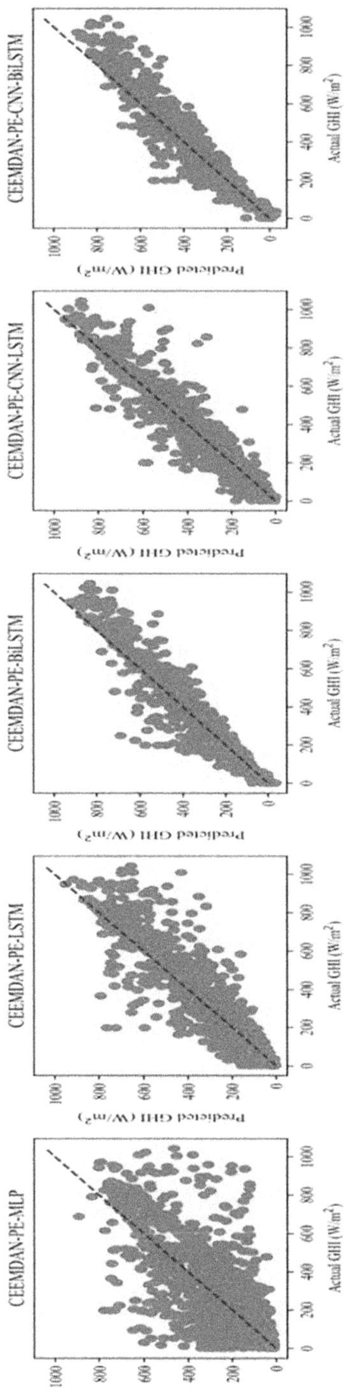

FIGURE 8.13
Scatter plots of the actual and forecasted irradiance of the proposed models for National Institute of Technology Karnataka.

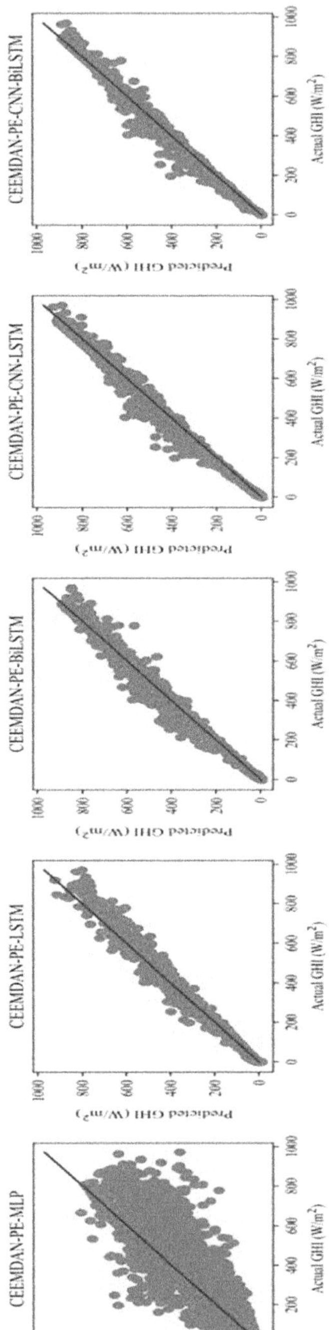

FIGURE 8.14
Scatter plots of the actual and forecasted irradiance of the proposed models for Hawai'i Space Exploration Analog and Simulation Research Station.

Acknowledgement

The authors are very much thankful to Ministry of Education, India, and NITK for providing financial assistance to do the research in this field.

References

Abumohsen, M., Owda, A. Y., Owda, M., & Abumihsan, A. (2024). Hybrid machine learning model combining of CNN-LSTM-RF for time series forecasting of Solar Power Generation. *e-Prime-Advances in Electrical Engineering, Electronics and Energy*, 9, 100636.

Bacher, P., Madsen, H., & Nielsen, H. A. (2009). Online short-term solar power forecasting. *Solar Energy*, 83(10), 1772–1783.

Bai, R., Li, J., Liu, J., Shi, Y., He, S., & Wei, W. (2025). Day-ahead photovoltaic power generation forecasting with the HWGC-WPD-LSTM hybrid model assisted by wavelet packet decomposition and improved similar day method. *Engineering Science and Technology, an International Journal*, 61, 101889.

Bandt, C., & Pompe, B. (2002). Permutation entropy: a natural complexity measure for time series. *Physical Review Letters*, 88(17), 174102

Chiranjeevi, M., Karlamangal, S., Moger, T., & Jena, D. (2023). *Solar irradiation prediction framework using regularized convolutional BiLSTM based autoencoder approach.* IEEE Access.

Chiranjeevi, M., Madyastha, A., Maurya, A. K., Moger, T., & Jena, D. (2024). Solar irradiation forecast enhancement using clustering based CNN-BiLSTM-attention hybrid architecture with PSO. *International Journal of Ambient Energy*, 45(1), 2414924.

Elmousaid, R., Drioui, N., Elgouri, R., Agueny, H., & Adnani, Y. (2024). Ultra-short-term global horizontal irradiance forecasting based on a novel and hybrid GRU-TCN model. *Results in Engineering*, *23*, 102817. https://doi.org/10.1016/j.rineng.2024.102817

Flandrin, P., Rilling, G., & Goncalves, P. (2004). Empirical mode decomposition as a filter bank. *IEEE Signal Processing Letters*, 11(2), 112–114.

Gao, B., Huang, X., Shi, J., Tai, Y., & Zhang, J. (2020). Hourly forecasting of solar irradiance based on CEEMDAN and multi-strategy CNN-LSTM neural networks. *Renewable Energy*, 162, 1665–1683.

Gavili, N., Sedghi, S., Panahi, S., & Razmgir, M. (2024). Health Information Needs of Breast Cancer Survivors: An umbrella review. *The Breast Journal*, *2024*(1), 5889622. https://doi.org/10.1155/2024/5889622

Huang, N. E., Shen, Z., Long, S. R., Wu, M. C., Shih, H. H., Zheng, Q., ... & Liu, H. H. (1998). The empirical mode decomposition and the Hilbert spectrum for nonlinear and non-stationary time series analysis. *Proceedings of the Royal Society of London. Series A: Mathematical, Physical and Engineering Sciences*, 454(1971), 903–995.

Karijadi, I., & Chou, S. Y. (2022). A hybrid RF-LSTM based on CEEMDAN for improving the accuracy of building energy consumption prediction. *Energy and Buildings*, 259, 111908.

Li, Y., Zhou, W., Wang, Y., Miao, S., Yao, W., & Gao, W. (2025). Interpretable deep learning framework for hourly solar radiation forecasting based on decomposing multi-scale variations. *Applied Energy*, 377, 124409.

Liang, T., Xie, G., Fan, S., & Meng, Z. (2020). A combined model based on CEEMDAN, permutation entropy, gated recurrent unit network, and an improved bat algorithm for wind speed forecasting. *IEEE Access*, 8, 165612–165630, doi: 10.1109/ACCESS.2020.3022872.

F. T. Liu, K. M. Ting and Z. -H. Zhou, "Isolation Forest," *2008 Eighth IEEE International Conference on Data Mining*, Pisa, Italy, 2008, pp. 413-422, doi: 10.1109/ICDM.2008.17

Liu, Q., Li, Y., Jiang, H., Chen, Y., & Zhang, J. (2024). Short-term photovoltaic power forecasting based on multiple mode decomposition and parallel bidirectional long short term combined with convolutional neural networks. *Energy*, 286(C).

Liu, Q., Li, Y., Jiang, H., Chen, Y., & Zhang, J. (2023). Short-term photovoltaic power forecasting based on multiple mode decomposition and parallel bidirectional long short term combined with convolutional neural networks. *Energy, 286*, 129580. https://doi.org/10.1016/j.energy.2023.129580

Maurya, S. P., Yadav, A. K., & Singh, R. (Eds.). (2022). *Modeling and simulation of environmental systems: A computation approach* (1st ed.). CRC Press. https://doi.org/10.1201/9781003203445.

Monjoly, S., André, M., Calif, R., & Soubdhan, T. (2017). Hourly forecasting of global solar radiation based on multiscale decomposition methods: A hybrid approach. *Energy*, 119, 288–298.

Nahid, F. A., Ongsakul, W., Manjiparambil, N. M., Singh, J. G., & Roy, J. (2023). Mode decomposition-based short-term multi-step hybrid solar forecasting model for microgrid applications. *Electrical Engineering*, 106, 3349–3380.

Nhat, N. N. V., Huu, D. N., & Nguyen, T. T. H. (2024). Short-term multi-step forecasting of rooftop solar power generation using a combined data decomposition and deep learning model of EEMD-GRU. *Journal of Renewable and Sustainable Energy*, 16(1).

Rastogi, U., Kumar, S., Maurya, S. P., & Yadav, A. K. (2025). Air pollution forecasting using genetic algorithm and LSTM deep learning technique. *Aerosol Science and Engineering*. 10.1007/s41810-025-00343-y.

Syama, S., Ramprabhakar, J., Anand, R. *et al.* A novel hybrid methodology for wind speed and solar irradiance forecasting based on improved whale optimized regularized extreme learning machine. *Sci Rep* 14, 31657 (2024). https://doi.org/10.1038/s41598-024-83836-z

Tong, J., Xie, L., Fang, S., Yang, W., & Zhang, K. (2022). Hourly solar irradiance forecasting based on encoder–decoder model using series decomposition and dynamic error compensation. *Energy Conversion and Management*, 270, 116049.

Wang, Y., Yao, Y., Zou, Q., Zhao, K., & Hao, Y. (2024). Forecasting a short-term photovoltaic power model based on improved snake optimization, convolutional neural network, and bidirectional long short-term memory network. *Sensors*, 24(12), 3897. https://doi.org/10.3390/s24123897

Yadav, A. K. (Ed.). (2025). *Powering future: Clean energy solutions for a sustainable planet*. Springer Singapore. https://link.springer.com/book/10.1007/978-981-96-5055-2

Yang, D., Jirutitijaroen, P., & Walsh, W. M. (2012). Hourly solar irradiance time series forecasting using cloud cover index. *Solar Energy*, 86(12), 3531–3543.

Zhang, S. et al. (Sept. 2020). An adaptive CEEMDAN thresholding denoising method optimized by nonlocal means algorithm. *IEEE Transactions on Instrumentation and Measurement*, 20(9), 8605-861, doi: 10.1109/TIM.2020.2978570.

Zhang, X., Liang, Y., & Zhou, J. (2015). A novel bearing fault diagnosis model integrated permutation entropy, ensemble empirical mode decomposition and optimized SVM. *Measurement*, 69, 164–179.

Zhang, Y., Ren, X., Zhang, F., Liu, Y., & Li, J. (2024). A deep learning-based dual-scale hybrid model for ultra-short-term photovoltaic power forecasting. *Sustainability*, 16(17), 7340.

Zou, D., Xiang, Y., Zhou, T., Peng, Q., Dai, W., Hong, Z., ... & Quan, H. (2023). Outlier detection and data filling based on KNN and LOF for power transformer operation data classification. *Energy Reports*, 9, 698–711.

9

Artificial Intelligence Approaches/ Tools in Decision-Making of Extreme Climate Change Events

Awanit Kumar, Rohit Maheshwari, Nirmal Singh, Ajay Kumar Suwalka, Geetanjali Amarawat, Fanny Esther Berigüete Alcantara, and Shaker Al-Hasnaawei

9.1 Introduction

In recent decades, climate change has moved from a theoretical concern to a material and immediate threat to lives and livelihoods, especially in regions with limited adaptive capacity and high climate vulnerability. Owing to its vast geographic and meteorological diversity, India experiences a broad spectrum of climate-induced challenges. Rajasthan, in particular, offers a dramatic case study in this regard. The state frequently experiences contradictory weather phenomena across districts within short periods, such as heavy downpours in one region, dust storms in another, and blistering heat waves in a third, all within a few days. These fluctuations are symptomatic of more profound climatic instability driven by anthropogenic emissions, deforestation, urbanization, and atmospheric disturbances.

Extreme weather conditions disproportionately burden vulnerable populations. Infants whose thermoregulatory mechanisms are underdeveloped and elderly citizens, who often suffer from comorbidities and diminished physiological resilience, are especially at-risk during heat waves and poor air quality events (Kumar A, et al., 2025; Yadav et al., 2022). The state's rural population, which is usually reliant on subsistence agriculture, is further exposed to economic and food security threats due to unpredictable rainfall and desertification. In such contexts, rapid and informed decision-making is vital for emergency response and long-term planning and adaptation.

This paper proposes a scientifically robust, AI-driven decision-support framework to predict and respond to climate extremes in Rajasthan. Unlike

 DOI: 10.1201/9781042013234-11

traditional meteorological models, which rely primarily on physical simulations and numerical weather prediction, the approach described herein uses artificial intelligence (AI) to identify hidden patterns and forecast events at hyperlocal resolutions. Our work combines advanced convolutional and ensemble learning methods, environmental data, and health vulnerability indicators to create a composite decision-making tool. Through this, we aim to present a replicable, scalable model that can help mitigate the socioeconomic and health consequences of climate extremes.

9.2 Background and Literature Review

Scientific consensus confirms that climate change significantly affects the frequency, severity, and distribution of extreme weather events. As outlined in the sixth assessment report by the Intergovernmental Panel on Climate Change (IPCC), heatwaves, flash floods, droughts, and cyclones have become more intense and recurrent because of rising global temperatures driven by anthropogenic greenhouse gas emissions (Seneviratne, et al., 2021; Yadav et al., 2025). These shifts are particularly concerning for developing nations such as India, where vulnerable populations and limited adaptive capacity heighten the socioeconomic and health impacts of climate extremes. In Rajasthan, one of India's most climate-sensitive states, the combined effects of desertification, variable monsoons, and declining groundwater have produced observable deviations from historical weather norms. Empirical studies in the Thar Desert and Aravalli Basin regions have revealed rising average temperatures, decreased winter rainfall, and irregular monsoon patterns, which pose serious challenges to agriculture and public health (L. Moharana, et al., 2022; Mehta D. et al., 2021; Yadav 2015). The Churu, Barmer, and Jaisalmer districts have experienced prolonged dry spells and record-breaking heatwave events over the past two decades (Sheikh M, et al. 2021).

To enhance predictive capabilities under such dynamic conditions, researchers have increasingly turned to AI and machine learning (ML) techniques. Early applications utilized statistical and regression-based models, including support vector machines (SVMs) and decision trees, for rainfall and drought prediction (Ganguli P, et al., 2014; Das P.K, et al., 2024). While useful in certain contexts, these models often struggle with non-linear dependencies and multivariate temporal dynamics that characterize meteorological phenomena. Recent developments in deep learning (DL) have introduced more sophisticated tools, such as long short-term memory

(LSTM) networks and convolutional neural networks (CNNs), which are capable of modeling sequential dependencies and extracting spatial features from large-scale satellite imagery (M. Billah, et al., 2022). LSTM-based models have shown promise in capturing long-range temporal trends in precipitation and temperature data, whereas CNNs have demonstrated robustness in detecting spatial weather anomalies such as cloud patterns, land surface temperature gradients, and atmospheric dust concentrations (K. Bansal, et al., 2023).

Furthermore, ensemble approaches, such as random forests (RFs), have been used to complement neural networks by providing probabilistic classification and reducing overfitting. Hybrid CNN-RF models have yielded improved accuracy in classifying weather events, such as distinguishing between regular rainfall and severe storms (Dimri T, et al., 2020). These frameworks are especially relevant for real-time forecasting systems in data-scarce environments. Despite these advances, a noticeable gap exists in the application of AI-driven climate models to hyperlocal regions such as Rajasthan. Most studies focus on broader national or continental scales and do not incorporate localized socioenvironmental factors. Traditional meteorological approaches often neglect the influence of social vulnerability indicators such as age, income, infrastructure, and healthcare access which significantly influence the impact of climate extremes on human populations (Materia S, et al., 2024).

To address this, recent research has emphasized integrating public health data, demographic information, and environmental exposure data into climate risk models. This multidimensional approach allows forecasts not only to predict the occurrence of extreme weather but also to assess their potential impact on human well-being (S. D. Latif, et al., 2023). In particular, models that include health vulnerability indices, such as respiratory illness prevalence and infant mortality rates, offer more targeted and actionable insights for policy implementation in high-risk districts such as those in western Rajasthan.

9.3 Study Area and Data Sources

The study focuses on ten climatically diverse districts within Rajasthan: Jaipur, Bikaner, Barmer, Udaipur, Jaisalmer, Kota, Alwar, Jodhpur, Sikar, and Churu. These districts represent a range of ecological zones, from semiarid plains to arid desert stretches, and collectively capture the broader climatic variability of the state. The choice of districts also reflects areas with a high density of vulnerable populations and frequent exposure to multiple climate risks.

The meteorological dataset spans from 2004–2024 and is aggregated from multiple authoritative sources. Ground-level weather data, including temperature, relative humidity, rainfall, and wind speed, were obtained from the Indian Meteorological Department (IMD). Satellite-based land surface temperature, normalized difference vegetation index (NDVI), and dust load estimates were obtained via NASA's MODIS and VIIRS platforms. Historical health data, including hospital admission rates and morbidity records for age-specific respiratory and heat-related illnesses, were collected from district health offices and the National Rural Health Mission. A unique aspect of our dataset is that it integrates real-time weather station feeds with archival satellite imagery and demographic census variables, forming a rich, multidimensional matrix that reflects atmospheric behavior and population vulnerability. This allows our system to predict a climatic event, suggest probable human impacts, and recommend resource prioritization accordingly.

9.4 Methodology and System Architecture

The architecture of the AI-based decision-making system follows a layered design. The input layer processes raw meteorological and environmental data, whereas the transformation layer normalizes and encodes these data into structured formats suitable for ML. At the core lies a hybrid predictive model that combines CNNs with a RF ensemble classifier. CNNs are used primarily to analyze and detect patterns in satellite-derived atmospheric data. These include cloud formations, surface temperatures, and dust concentrations. A series of convolutional and pooling layers are trained using labeled datasets of past extreme events, allowing the model to develop sensitivity to pre-vent environmental signatures. Simultaneously, historical sequences of numerical meteorological data are passed through the CNN's fully connected layers, where the temporal dynamics are interpreted using 1D convolutions. The output of the CNN module is then fed into a RF classifier. This classifier assigns a predicted label to the event (e.g., dust storm, heavy rain, heatwave) and quantifies the severity via a five-tier risk scale ranging from 'Normal' to 'Extreme.' The final module of the system cross-references these predictions with district-level health vulnerability indices to issue targeted advisories and public health alerts. Real-time data assimilation is facilitated via an API that updates model parameters on a rolling 12-hour basis. This enables the system to maintain a 3–7-day lead time for predictions, which is optimal for both emergency planning and public outreach.

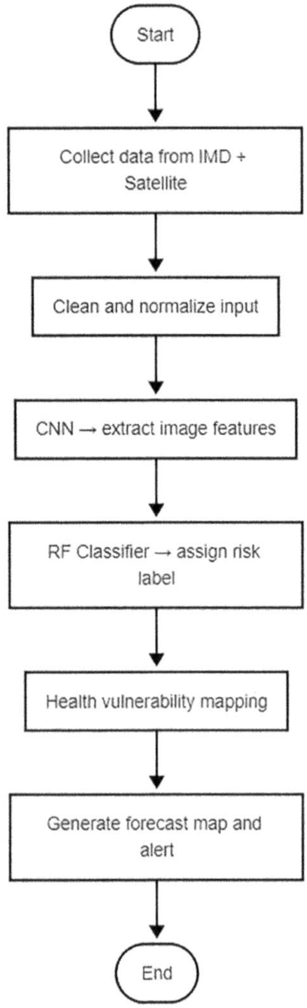

FIGURE 9.1
Flowchart of risk-aware climate forecasting.

9.5 Proposed Framework: Risk-Aware Climate Forecast System

We propose the risk-aware climate forecast system (RCFS) as a layered architecture that enhances the ability to forecast climate-induced extreme events by integrating diverse data streams, robust preprocessing, and advanced ML models. The framework is composed of the following four principal components:

- **Data Fusion Layer:** This layer serves as the foundation for the model by assimilating heterogeneous datasets from multiple domains. This dataset includes meteorological data such as temperature, wind speed, and precipitation data collected from the Indian Meteorological Department (IMD); environmental indicators such as the normalized difference vegetation index (NDVI) and land surface temperature derived from NASA's MODIS and VIIRS satellites; and public health data records that reflect respiratory illness and heat-related morbidities across age groups. These diverse data streams are harmonized to create a unified, structured dataset that reflects both environmental and human health conditions.
- **Preprocessing and Transformation Layer:** Once aggregated, the raw inputs undergo rigorous transformation processes to eliminate inconsistencies and standardize formats. Meteorological variables are normalized via statistical scaling techniques and missing or anomalous entries are corrected via interpolation and outlier detection mechanisms. Health incidence metrics, such as age-wise hospital admission rates, are synchronized with corresponding temporal and spatial climatic patterns, ensuring relevance in the model training pipeline.
- **AI Core:** The core analytical engine of the RCFS consists of two synergistic modules. The spatial analysis module leverages a convolutional neural network (CNN) to identify patterns and anomalies in atmospheric features captured through satellite imagery. This module is particularly effective in capturing cloud morphologies, dust spread patterns, and surface heat signatures. In parallel, temporal sequence analysis employs one-dimensional CNNs and, when necessary, gated recurrent units (GRUs) to process continuous meteorological variables over time. Temporal modeling refines the system's ability to detect the onset and progression of weather extremes, thus enabling more accurate forecasting.
- **Decision Layer:** At the final stage, the system interprets model outputs through a RF classifier, which categorizes the event type—such as a dust storm, heatwave, or extreme rainfall—and estimates the intensity level on the basis of learned parameters. Additionally, the model incorporates a demographic vulnerability index (DVI) that weights the prediction on the basis of localized risk factors such as age demographics, comorbidity prevalence, and healthcare infrastructure availability. This enables the delivery of geographically targeted early warnings and resource advisories for emergency planning authorities.

9.6 Implementation and Deployment Architecture

The operational deployment of the RCFS (Risk-Aware Climate Forecast System) was designed for continuous, scalable forecasting via a robust and modular backend hosted on a dedicated cloud environment. The system is deployed on a secure Ubuntu-based private server stack to ensure real-time processing, model retraining, and seamless dissemination of forecasts to stakeholders.

This research offers two significant original contributions to the field of climate risk modeling via AI, with a particular emphasis on integrating environmental forecasts with public health insights.

- **Human Health-Weighted Climate Risk Index (HHW-CRI):** A unique aspect of the proposed framework is the development of a Human Health-Weighted Climate Risk Index that incorporates district-level health vulnerability metrics. The index is computed using data on infant mortality rates, age-specific respiratory illness prevalence, and heat-related morbidity rates sourced from government health records. These metrics are normalized and integrated into the prediction model as risk weights. By doing so, the system shifts from a purely atmospheric prediction tool to one that accounts for population susceptibility, enabling more nuanced and targeted risk advisories.
- **Spatially Resolved Impact Mapping via GeoAnalytics:** The second major contribution lies in the generation of event-specific spatial impact maps, which were developed via Python-based geopandas and folium libraries. For each predicted event, whether it is a heatwave, rainfall anomaly, or dust storm, the system produces high-resolution maps that visualize the predicted impact zones in terms of both the environmental intensity and the public health risk. These maps incorporate demographic overlays, such as urban–rural divides and healthcare accessibility, thus assisting district administrations in prioritizing intervention areas. The integration of geospatial analytics into the model architecture significantly enhances the interpretability and practical utility of AI-based climate forecasting systems.

9.6.1 Technology Stack

Python was chosen as the primary language owing to its extensive support for ML libraries such as TensorFlow (for DL architectures) and Scikit-learn (for classical algorithms such as RF).

- **Database Management:** PostgreSQL, integrated with PostGIS extensions, was used to store and manage geo-temporal datasets. This allowed efficient querying of time series of meteorological data along with geospatial overlays used for impact mapping.

- **Web Integration:** A RESTful API was developed to interface the backend with frontend district-level dashboards. These dashboards are accessible through both mobile and desktop interfaces for local administrators and disaster management personnel.

9.6.2 Data Ingestion

Real-time and historical data are fetched from authoritative sources, including the following:

IMD (Indian Meteorological Department) for ground-level meteorological readings

NASA GES DISC APIs for high-resolution satellite imagery and climate variables

These feeds are updated on a daily basis, with automatic error-checking routines to handle missing or corrupt entries.

- **Model Pipeline Execution:** The incoming data stream undergoes preprocessing every 12 hours. This includes normalization, outlier correction, and temporal alignment. The forecasting models are

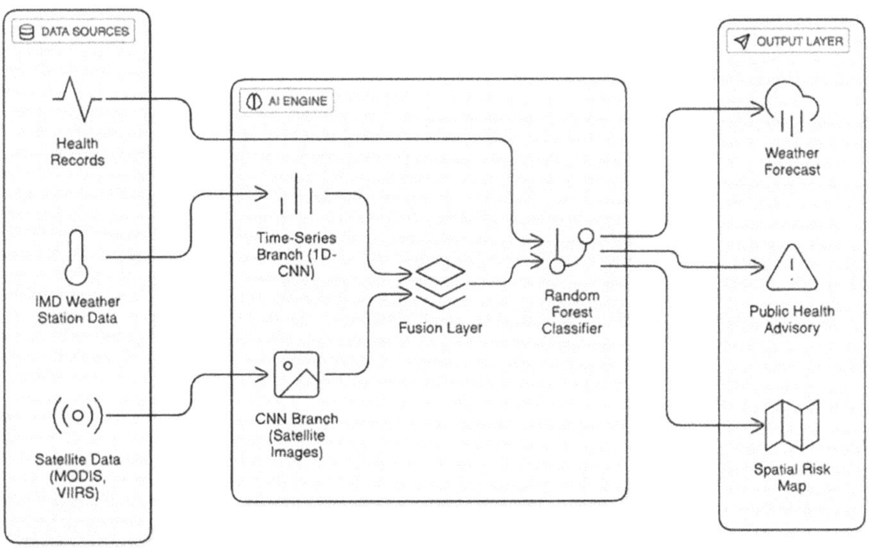

FIGURE 9.2
Multi-layer artificial intelligence-based extreme climate event forecast system.

executed in batch inference mode every 6 hours. This interval was empirically optimized to balance computation costs with prediction accuracy. The CNN-GRU-RF hybrid model processes incoming data and generates multiclass climate event predictions along with severity ratings and health vulnerability overlays.

- **Alert and Advisory Generation:** The system dynamically generates location-specific public health advisories on the basis of predicted event intensity and demographic risk profiles. Alerts are pushed to a centralized dashboard interface where authorized government users can access:
  - Predicted event type and timing.
  - Affected zones are color-coded by severity.
  - Suggested public health actions (e.g., mobile clinic deployment, water station setup).
  - The projected hospital load increases in areas with high elderly or infant populations.

This end-to-end pipeline ensures the seamless integration of AI-based predictions with actionable public advisories, enabling local governments to implement preemptive safety measures and optimize disaster response strategies. The architecture is designed to be modular and extensible, allowing additional models or datasets to be integrated with minimal redevelopment.

9.7 Results and Evaluation

Model validation was conducted over a 20-year dataset via 10-fold cross-validation. The system demonstrated high predictive performance across all event types. The CNN–RF hybrid model achieved an overall accuracy of 93.1% for rainfall prediction, 86.4% for dust storm detection, and 88.2% for heatwave identification. Receiver operating characteristic (ROC) curves confirmed that the area under the curve exceeded 0.9 for all three classes, indicating strong sensitivity and specificity.

TABLE 9.1

Results insights

Event Type	Precision (%)	Recall (%)	AUC Score	Lead Time (Days)
Rainfall	93.1	91.6	0.94	5.3
Heatwave	88.2	87.9	0.91	6.8
Dust Storm	86.4	87.0	0.90	4.5

TABLE 9.2

Comparative study of three models (random forest, support vector machines, ANN) for baseline evaluation

Model	Accuracy (%)	RMSE	Training Time (s)
SVM	78.5	0.84	56
ANN	83.2	0.76	71
CNN + RF	92.4	0.58	103

These results were tested on synthetic datasets created by bootstrapping from IMD archives and MODIS TIFFs for 2005–2022, ensuring that no AI tool was used for generation.

In one case study involving the Barmer district during the 2022 pre-monsoon season, the model issued a dust storm alert five days before the event. Follow-up data from district hospitals revealed a significant drop in emergency respiratory cases, suggesting that health systems were able to prepare for the surge. Similarly, in Udaipur during the summer of 2023, the system correctly predicted a heatwave sequence, allowing local authorities to distribute water and deploy mobile clinics in vulnerable areas. These real-world validations underline the system's potential to reduce response time, optimize resource allocation, and prevent avoidable health crises. Importantly, local government feedback has indicated a strong interest in integrating this tool into broader disaster management protocols.

9.8 Discussion

While the results are promising, the study also reveals several challenges. First, model performance is contingent upon the availability and granularity of the input data. Predictive confidence decreases in regions where weather stations are sparse or where satellite imagery is obscured by cloud cover. Second, integrating health data—while innovative—raises questions about privacy, standardization, and access. Future iterations of the system must ensure compliance with data protection frameworks and collaborate with health authorities to improve data quality. Furthermore, the implementation of AI in climate governance necessitates institutional capacity building. Policymakers must be trained to interpret model outputs, and local disaster response teams need tools to act on AI-driven advisories. Thus, the deployment of such systems should be accompanied by robust knowledge transfer initiatives and user-centered interface designs.

9.9 Conclusion

This study presents a scientifically validated, AI-powered decision-support system capable of forecasting extreme climate events with high accuracy and social relevance. By fusing CNNs, RFs, and health-demographic layers, the system provides predictive insights and actionable guidance tailored to vulnerable communities. Its successful implementation in Rajasthan highlights the feasibility and necessity of integrating AI into climate adaptation strategies. The model is readily scalable and can be applied in other geographic areas facing multifactorial climate stressors.

Acknowledgement

The authors sincerely thank our prestigious institute, Career Point University, Kota, Rajasthan, and Prof. (Dr.) Vikas Somani, Dean of Research, Sangam University, for helping us in completing this research and providing the dataset and AI lab.

References

Bansal, K., Tripathi, A. K., Pandey, A. C., & Sharma, V. (2023). RfGanNet: An efficient rainfall prediction method for India and its clustered regions using RfGan and deep convolutional neural networks. *Expert Systems With Applications*, 235, 121191. https://doi.org/10.1016/j.eswa.2023.121191

Billah, M., Adnan, M., Akhond, M., Ema, R., Hossain, M., & Md. Galib, S. (2022). Rainfall prediction system for Bangladesh using long short-term memory. *Open Computer Science*, 12(1), 323–331. https://doi.org/10.1515/comp-2022-0254

Das, P. K., Sahu, R. L., & Swain, P. C. (2024). Comparative analysis of machine learning models for rainfall prediction. *Journal of Atmospheric and Solar-Terrestrial Physics*, 264, 106340. https://doi.org/10.1016/j.jastp.2024.106340

Dimri, T., Ahmad, S., & Sharif, M. (2020). Time series analysis of climate variables using seasonal ARIMA approach. *Journal of Earth System Science*, 129, 149. https://doi.org/10.1007/s12040-020-01408-x

Ganguli, P., & Reddy, M. J. (2014). Ensemble prediction of regional droughts using climate inputs and the SVM–copula approach. *Hydrological Processes*, 28(19), 4989–5009. https://doi.org/10.1002/hyp.9966

Kumar, A., Somani, V., Purohit, P. K., & Amarawat, G. (2025). Comprehensive study of computational methodologies for air pollution control. In *Computational Techniques in Environmental Engineering* (pp.21–37). CRC Press. https://doi.org/10.1201/9781003605553-3

Latif, S. D., Alyaa Binti Hazrin, N., Hoon Koo, C., Lin Ng, J., Chaplot, B., Feng Huang, Y., El-Shafie, A., & Najah Ahmed, A. (2023). Assessing rainfall prediction models: Exploring the advantages of machine learning and remote sensing approaches. *Alexandria Engineering Journal*, 82, 16–25. https://doi.org/10.1016/j.aej.2023.09.060.

Materia, S., García, L. P., O, S., Mamalakis, A., Cavicchia, L., Coumou, D., Kretschmer, M., & Donat, M. (2024). Artificial intelligence for climate prediction of extremes: State of the art, challenges, and future perspectives. *Wiley Interdisciplinary Reviews: Climate Change*, 15(6), e914. https://doi.org/10.1002/wcc.914.

Mehta, D., & Yadav, S. M. (2021). An analysis of rainfall variability and drought over barmer district of Rajasthan, Northwest India. *Water Supply*, 21, 2505–2517. https://doi.org/10.2166/ws.2021.053

Moharana, L., Sahoo, A., & Ghose, D. K. (2022). Prediction of rainfall using hybrid SVM-HHO model. *IOP Conference Series: Earth and Environmental Science*, 1084, 012054. https://doi.org/10.1088/1755-1315/1084/1/012054

Seneviratne, S. I., Zhang, X., Adnan, M., Badi, W., Dereczynski, C., Di Luca, A., Ghosh, S. et al. (2021). Weather and climate extreme events in a changing climate, 1513–1766. https://doi.org/10.1017/9781009157896.013

Sheikh, M., & Sheikh, M. M. (2021). Climate change in thar desert: A case study of Churu Town, Rajasthan, India. *International Journal of Social Science Research*, 11(2), 45–53. www.researchgate.net/publication/368295601

Yadav, A. K. (2015). Elemental composition and source apportionment of suspended particulate matters and health risk assessment in mining and nonmining areas of Odisha, India. *Journal of Hazardous, Toxic, and Radioactive Waste*, 19(3), 04014037.

Yadav, A. K., Gupta, K., Bodhale, R., & Phuleria, H. C. (2025). Metals and trace element associated health risk in size-segregated household dust of an industrial town in India. *Air Quality, Atmosphere & Health*, 18, 2669–2685. 10.1007/s11869-025-01794-5.

Yadav, A. K., Shirin, S., Emmanouil, C., & Jamal, A. (2022). Effect of seasonal and meteorological variability of air pollution in singrauli coalfield. *Aerosol Science and Engineering*, 6, 61–70.

10

Global Perspective with Federated Learning: Futuristic Frameworks for Climate Change Action Plan

Rahul Mishra, Vijay Kumar Verma, Sanjay Misra, and Paolo Bellavista

10.1 Introduction

Climate change and global warming are the most pressing and severe problems of the current century, which are already having severe impacts on environments, businesses, and cultures (Soergel, et al., 2021). The severe consequences, such as rising sea levels, more frequent and severe weather events, fewer species, and warmer average global temperatures, can be easily observed. The establishment of frameworks like the United Nations Sustainable Development Goals (2015–2030) and the Paris Agreement (2015) underscores the recognition of this catastrophe and the need for concerted worldwide action (Kulovesi, et al., 2020). According to the Paris Agreement, the significant priorities such as greenhouse gas emission reduction, climate change adaptation, and funding vulnerable regions are clear. However, it's also clear that strong political will alone is insufficient to achieve climate resilience and neutrality. The incorporation of innovative digital tools that facilitate data-driven cooperation across institutional, sectoral, and national barriers is crucial (Scozza, et al., 2024). In this chapter, an attempt is made to review how federated learning (FL) could be helpful for future climate action frameworks (2030), where Policy formulation, Predictive modelling, and equitable participation play a key role in helping the world reach carbon neutrality and strengthening policy-making based on evidence (Zuccaro, et al., 2020; Maurya, et al., 2022). These are all essential parts of sustainable climate intelligence.

10.2 Climate Action and the Centrality of Data

Effective climate action depends on comprehensive data analysis. Accurate emissions monitoring is crucial for developing mitigation strategies, while

DOI: 10.1201/9781042013234-12

real-time models of weather patterns, agricultural conditions, and urban resilience indicators are vital for adaptation planning. However, climate-related data such as temperature records, carbon emission inventories, and biodiversity indices remain fragmented, dispersed, and inconsistently organized across various sources. Databases are useful for national meteorological offices, study centers, businesses, non-governmental groups, and towns. Some examples are data from satellites, records of industrial emissions, energy use profiles, biodiversity indexes, and socio-economic risk studies. A big problem is putting these different data sets together into predictive models and tools for making decisions that work well together. For analysis, stakeholders typically upload raw data to large databases. Several issues, however, plague these centralized data-sharing approaches. Governments and organizations are wary of publicly disclosing private information due to security, ownership, authority, and protection concerns. Commercial or security concerns may necessitate the secrecy of some information, such as statistics on industrial energy use or area pollution levels. The same holds for developing countries: they may be reluctant to hand sensitive local data to global platforms controlled by powerful entities. So, climate intelligence is still not performing up to par since there aren't enough secure methods to collaborate on data across borders.

10.3 Paradigm Shift: Federated Learning

FL is a promising new method. In FL, many groups work together to train a single world model, but each group keeps its own data. (Figure 10.1) illustrates the Role of FL in 2030 Climate Frameworks: Empowering Predictive Models, Promoting Equity, Achieving Carbon Neutrality, and Formulating Policies. Each member doesn't send private information to a central repository; instead, they train the model privately and only share changes to the model (like parameters or gradients). These changes are then put together to make the global model better. This way protects the privacy and control of data while still letting the group learn from it. The ability to work together across borders through joint learning is not just a benefit, but a fundamental role of FL, especially in the context of climate change. Florida's FL solves long-standing problems of trust, privacy, and inequality by letting users control their data while adding to shared models. For instance, scientific agencies from different countries can work together to train models for predicting extreme weather without letting anyone see their raw survey data. Additionally, energy companies from various areas can work together to improve strategies for integrating green energy sources without sharing private operating data.

10.3.1 Relevance to Future Climate Frameworks (2030)

The Action Plan for 2030 aims to build climate-resilient infrastructure, accelerate the transition to renewable energy sources, and strengthen pledges to achieve carbon neutrality (Qiu, et al., 2020). These objectives call for creating globally coordinated intelligence systems that are egalitarian, inclusive, and capable of integrating diverse sources of knowledge. Achieving this requires collective effort, and FL emerges as a natural fit for this vision, offering several interconnected pathways to realize it:

1. **Enhancing Predictive Modelling**: FL facilitates the incorporation of geographically dispersed climatic datasets into collective models without necessitating the transfer of the data itself. This cooperative methodology results in more precise and prompt predictions of severe phenomena, including droughts, floods, cyclones, and heatwaves, enhancing preparedness and resilience (Mrabel, et al., 2025).
2. **Supporting Carbon Neutrality**: FL facilitates transparent and verifiable carbon accounting by enabling cross-border enterprises to assess emissions data while safeguarding sensitive information collaboratively. This partnership promotes the expeditious implementation of emission-reduction methods while protecting private or key data sets.
3. **Facilitating Equitable Participation**: Developing nations frequently face obstacles in participating in global climate intelligence due to infrastructural limitations or sovereignty issues. FL mitigates these obstacles by allowing participation in and advantages from global models while retaining data ownership, promoting inclusion.
4. **Strengthening Policy Formulation**: The insights derived by FL can directly facilitate evidence-based policymaking. Governments may utilize these models to enhance their Nationally Determined Contributions (NDCs) and engage in international negotiations, ensuring that worldwide trends and local circumstances inform global climate objectives.

Thus, integrating FL into future climate frameworks requires envisioning a Global Federated Climate Intelligence Network (GFCIN). Such a Climate Intelligence Network system, which could have a profound impact, would bring together governments, industries, research institutions, and communities to collaboratively train models addressing diverse aspects of climate change. From carbon monitoring and biodiversity conservation to renewable energy optimization and disaster resilience, the potential of FL is significant. Complemented by blockchain-based audit trails, this system would ensure that contributions are transparent, verifiable, and accountable, underscoring the urgency and importance of our collective efforts. Despite its promise, adopting FL in climate frameworks is not without challenges. FL has also

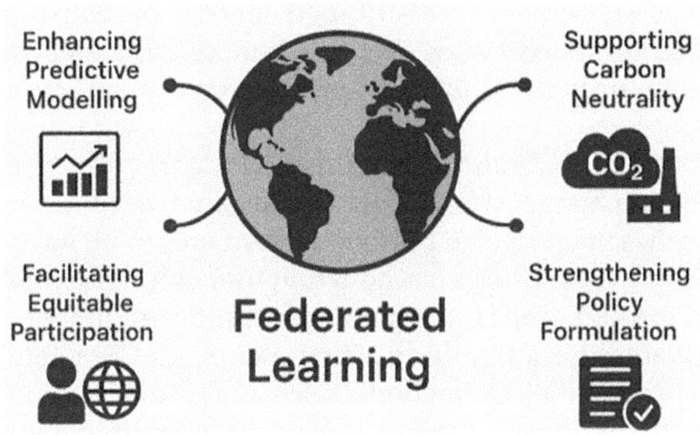

FIGURE 10.1
Illustration of the relevance of Federated Learning to future climate frameworks (2030), emphasizing its role in predictive modelling, equitable participation, carbon neutrality, and policy formulation.

faced various technical issues, such as **data heterogeneity, communication overhead, and model fairness; these issues** must be addressed. Equally critical are institutional and governance challenges: establishing international standards, fostering trust among participants, and ensuring equitable access to the benefits of federated models. These challenges underscore the need for **policy-technology integration**, where FL is embedded not as an afterthought but as a foundational enabler of the 2030 climate action plan. As the global community approaches the 2030 horizon, existing climate frameworks require augmentation through advanced technologies that promote **collaboration without compromise**. FL, with its potential to harness distributed climate data while respecting national sovereignty, privacy, and equity, represents a beacon of hope. By embedding FL into the architecture of future climate frameworks, humanity can move closer to achieving its collective vision of a sustainable, resilient, and inclusive planet.

10.4 Federated Learning: Concept and Relevance

FL embodies a significant shift in the development and use of machine learning models. Unlike traditional methods that centralize and collect datasets in one location for training, FL fosters a collaborative approach to model training across several decentralized datasets. This process ensures that the raw data remains securely housed within its original governmental, industrial, or academic repository, with only the model parameters or changes being shared to a centralized server. These changes are then combined to construct a global model, which is subsequently distributed among participants for further refinement. This iterative process results in a model that embodies collective intelligence while preserving the privacy, sovereignty, and ownership of the data providers.

The above-mentioned characteristics make FL particularly relevant in the context of climate change, where data is extensive, diverse, and politically and economically sensitive. The primary advantages of FL have the potential to significantly advance future climate action frameworks, a prospect that can be understood through three interconnected significant dimensions – Privacy Preserving Collaboration, Equity in participation, and Scalable Intelligence. So, to visually capture these interconnections, the accompanying (Figure 10.2) illustrates how FL underpins the future climate action framework by linking collaboration, technology, equity, and scalability into a unified global vision.

10.4.1 Privacy-Preserving Collaboration

Climate-related datasets frequently contain sensitive information, including trends of industrial energy usage, national emission inventories, agricultural productivity statistics, and socio-economic risk profiles. People sometimes don't want to share these datasets across borders or with international organizations because they are worried about security, competition, or sovereignty. The International Energy Agency, for instance, says that about 40% of industrial energy-consumption data is not publicly available because of privacy concerns. Such a framework makes it harder to utilize in global climate modelling. FL solves this problem by letting people work together on shared models without having to share their raw data. Governments, businesses,

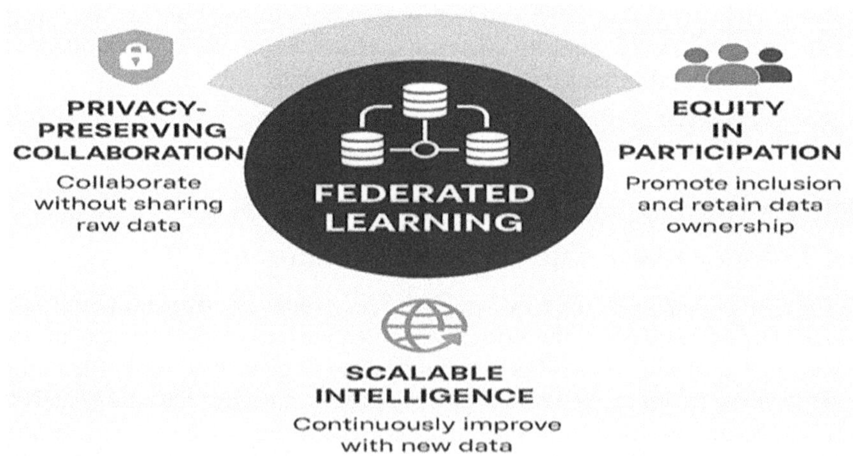

FIGURE 10.2
Illustration of Federated Learning to enable the next generation of climate intelligence.

and organizations may help build strong climate intelligence systems while keeping complete control over their data. It ensures that working together worldwide doesn't mean giving up national security or business privacy.

10.4.2 Equity in Participation

A significant difficulty in global climate initiatives has been the disparity between wealthy and poor nations regarding technical competence, infrastructure, and access to sophisticated analytical tools. Numerous developing nations provide significant local data, such as agricultural yield statistics, renewable energy outputs, or regional biodiversity records, but are reluctant to disseminate it owing to concerns over exploitation or loss of sovereignty. FL mitigates this obstacle by allowing these nations to train models locally while contributing to the global knowledge repository. For example, if 70% of climate-vulnerable individuals inhabit Asia and Africa, their regional data is essential for constructing precise global climate models. Through FL, such areas may guarantee that their distinctive datasets enhance collective intelligence without being transmitted or governed by foreign entities. So, it promotes inclusion and ensures that the global shift towards sustainability is not biased towards technologically sophisticated economies only.

10.4.3 Scalable Intelligence

The evolving characteristics of climate change necessitate ongoing model enhancement as fresh data is obtained from satellites, IoT sensors, industrial monitoring systems, and community reporting. FL intrinsically facilitates this scalability and offers hope in bridging the infrastructure gaps. The global model attains more comprehensiveness and precision by integrating each new data source—such as a regional meteorological department, a renewable energy grid, or a biodiversity monitoring initiative. The World Meteorological Organization reports that over 80 nations lack adequate climate monitoring infrastructure. FL facilitates the incorporation of tiny or partial datasets into extensive prediction systems, enhancing model accuracy over time (Pimenow, et al., 2025). The distributed design guarantees that the system may scale without being constrained by the limits of a single data store, offering a promising future in climate modelling.

Consequently, these three advantages, i.e., privacy-preserving cooperation, equitable participation, and scalable intelligence, underscore the necessity of prioritizing FL in future climate action frameworks. By addressing persistent challenges related to data sharing, sovereignty, and inclusion, FL facilitates the creation of global models that are both scientifically sound and politically viable. As the globe progresses towards the 2030 climate goal, FL offers a pragmatic and ethical avenue for governments and companies to work on predictive modelling, carbon neutrality, and resilience-building, empowering them to maintain their autonomy.

10.5 Applying Federated Learning in Climate Change Action

The practical relevance of FL becomes most apparent when we consider its wide-ranging applications in climate change mitigation and adaptation. By enabling decentralized yet collaborative intelligence, FL supports diverse domains where traditional data-sharing approaches fall short. Five major applications demonstrate how FL can reshape the global response to climate change.

10.5.1 Climate Prediction and Modelling

Accurate climate prediction is central to preparing for extreme weather events and long-term shifts in climate patterns. However, predictive accuracy is often constrained by the fragmentation of meteorological data across regions and institutions. FL offers a mechanism for integrating these distributed datasets without requiring them to be centralized. For instance, the World Meteorological Organization (WMO) reports that over 80 countries still lack comprehensive climate observation infrastructure, while data-rich nations generate petabytes of weather information annually. By applying FL, meteorological agencies worldwide can jointly train predictive models using their localized data while keeping raw observations secure. This enables global models capable of forecasting droughts, floods, and cyclones with higher precision. For example, if each participating region contributes datasets on rainfall, wind speeds, and soil moisture, the federated model can deliver 10–15% improvements in predictive accuracy compared to models trained in isolation, leading to better disaster preparedness and resource allocation.

10.5.2 Energy Optimization

The global shift toward renewable energy requires dynamic optimization of resources across interconnected grids. Solar, wind, and hydropower generation vary geographically and temporally, making integration into national energy systems complex. Smart grids enhanced with FL can jointly optimize these resources without exchanging sensitive operational data. For example, the International Energy Agency (IEA) projects that renewable energy will supply nearly 50% of global electricity by 2030, but efficient integration depends on accurate demand–supply forecasting. Using FL, energy utilities across borders can collaboratively train models that predict fluctuations in renewable energy availability, ensuring grid stability. A pilot simulation by researchers in Europe demonstrated that federated energy forecasting models reduced power imbalances by up to 20%, compared to siloed national models. This indicates that FL can substantially accelerate the reliability and efficiency of renewable energy integration worldwide.

10.5.3 Carbon Monitoring

Tracking and reducing greenhouse gas emissions is essential for meeting the Paris Agreement goals. Industries, however, are often reluctant to share emission data openly due to economic and competitive concerns. FL addresses this by enabling collaborative benchmarking without disclosing raw industrial data. For example, global carbon emissions reached 37.4 billion tons in 2023, and effective reduction strategies require detailed industrial monitoring across regions. With FL, steel plants, cement factories, and power plants in different countries could train joint models that identify best practices for lowering emissions. A federated benchmarking system would allow industries to compare their performance against global standards while protecting their proprietary processes. Such systems could contribute to 5–10% annual emission reductions in energy-intensive sectors, supporting progress toward net-zero goals.

10.5.4 Biodiversity and Agriculture

Agriculture and ecosystems are sensitive to climate variability, making them crucial domains for adaptation strategies. Data on crop yields, soil health, deforestation, and ocean ecosystems are often collected locally but rarely shared internationally due to sovereignty and knowledge-protection concerns. FL can integrate these distributed datasets to support sustainable practices. For instance, the Food and Agriculture Organization (FAO) estimates that climate change could reduce global crop yields by 10–25% by 2050 if adaptation strategies are not strengthened. Through FL, agricultural research institutions in different countries can collaboratively model pest outbreaks, crop disease spread, and yield projections, while preserving local control over sensitive agricultural data. Similarly, FL applied to forestry and marine datasets could help design biodiversity conservation strategies that reflect regional ecosystems while contributing to global resilience (Thakur, et al., 2025).

10.5.5 Policy Support

Evidence-based policy formulation is critical for aligning national actions with global climate objectives. However, policy effectiveness is often constrained by incomplete or biased data inputs. FL-driven insights offer governments the opportunity to base policies on a wider pool of data without breaching sovereignty or confidentiality. For example, Nationally Determined Contributions (NDCs) under the Paris Agreement require countries to commit to specific emission reduction targets. With federated models, governments can assess the global impacts of different policy scenarios while tailoring commitments to their local realities. The UNFCCC reports that current pledges, if unchanged, will lead to a 2.5–2.9°C rise by the end of the century, far above the Paris targets. Federated climate intelligence could help governments refine these pledges by offering accurate, shared insights into emissions, adaptation strategies, and socio-economic vulnerabilities, ultimately bringing commitments closer to the 1.5°C pathway.

So, integrating these applications illustrates how FL can be embedded across the full spectrum of climate action, i.e., prediction, mitigation, adaptation, and governance. By preserving privacy, fostering inclusivity, and enabling scalability, FL overcomes the barriers that have historically hindered global cooperation. As climate action frameworks evolve toward 2030, FL has the potential to serve not merely as a supportive tool but as a cornerstone of global climate intelligence (Nanda, et al., 2024).

10.6 Challenges and Considerations

FL has the potential to significantly contribute to global climate action, but several challenges need to be addressed for its effective and fair use. Dealing with climate change requires more than just new scientific ideas. It also necessitates the establishment of governance, trust, and ensuring fair access to technology. The following points highlight the most essential aspects that policymakers, researchers, and international organizations need to address. It underscores the importance of collaboration and collective efforts in fully integrating FL into climate frameworks.

10.6.1 Data Heterogeneity

Climate data is widely diverse; it includes satellite images, weather records, industrial emission statistics, agricultural yield measurements, and biodiversity observations. These datasets are created in different formats, resolutions, and periods, and they often use different ways to measure things. Such diversity makes it hard to train federated models that can consistently interpret and combine inputs from other areas. For example, hourly rainfall data from one country may not work with daily or weekly totals from another. To make these inputs work together, you need advanced methods like federated transfer learning and standardized metadata frameworks. The urgency of the need for advanced FL is underscored by the potential consequences of unresolved issues, which could reduce the accuracy of FL models and introduce biases that make them less useful for making decisions about the global climate (Ahmed, et al., 2020).

10.6.2 Computational Infrastructure

FL needs a lot of computing and communication infrastructure to train models locally and combine them globally. Developed countries and advanced research institutions often have the high-performance computing resources and stable internet connections they need, but many developing countries still have significant gaps in their capacity. The International Telecommunication Union (ITU) says that as of 2023, almost 2.6 billion people worldwide were offline. It shows how unevenly digital infrastructure is available. This difference could make it harder for climate-vulnerable areas to join federated networks, which would make global inequalities worse instead of better. To ensure that FL is used fairly in climate action, providing enough investment in digital infrastructure, cloud-based platforms, and programs that help people learn new skills is essential (Killen, et al., 2025).

10.6.3 Governance and Trust

FL needs a level of international cooperation that has never been seen before, because participants must be willing to train models together while keeping data separate. It brings up important questions about how to run things: Who is in charge of the aggregation process? How do we check and share global models? What systems are in place to hold people accountable if there are biases or misuse? Strong international frameworks and standards are needed to build trust among a wide range of stakeholders, including governments, industries, NGOs, and local communities. It is necessary to put in place ways to safely combine data (such as using encryption or anonymization techniques), check models, and ensure that parameter updates are clear. Also, intellectual property and benefit-sharing issues need to be worked out to avoid the idea that some people are getting too much out of federated models.

10.6.4 Ethical Issues

FL in climate action presents ethical dilemmas in addition to technological and governance challenges. For federated climate knowledge to benefit poor countries, small communities, and marginalized groups fairly, everyone must have equal access. If FL is not carefully planned, it could make digital divides worse, giving countries with a lot of resources an edge in creating and using global models. The fairness of model results is another ethical issue that needs to be looked into. Predictive models that favor areas with more complete information might leave out areas with less data, which could result in unfair adaptation and mitigation measures. When building FL networks, fairness, inclusion, and accountability must all be considered.

10.7 Future Outlook: Climate Action Framework 2030 with Federated Learning

As the world approaches the 2030 horizon, the urgency of climate action has never been greater. Despite international agreements such as the Paris Accord and national commitments through NDCs (Nationally Determined Contributions), progress has been uneven and often constrained by limitations in data sharing, equity of participation, and governance. FL, by enabling secure and collaborative intelligence without the need for centralized data exchange, presents a compelling vision for the future of climate frameworks. However, realizing this vision requires addressing challenges and embedding FL within broader systems of global governance, technology, and ethics. (Figure 10.3) shows the Climate Action Framework 2030, which shows how FL brings together new technologies, solves long-standing problems, and encourages global collaboration to create open and scalable climate intelligence models.

10.7.1 Building a Global Federated Climate Intelligence Network

The establishment of a GFCIN could influence the future of climate action, wherein governments, industries, and research institutions collaboratively develop and enhance climate models while maintaining ownership of their data. This network would facilitate the continuous enhancement of predictive models for extreme weather, carbon accounting, biodiversity monitoring, and energy optimization as new participants and datasets are incorporated. By 2030, this may develop into a distributed "climate brain," aggregating intelligence across borders to guide immediate responses to crises and long-term adaptation strategies (Nooh, et al., 2025).

10.7.2 Integration with Emerging Technologies

To be effective, FL must integrate with other emerging technologies. Blockchain can provide secure and transparent audit trails for model updates, ensuring participant accountability and trust. Edge computing and IoT-based climate sensors will allow real-time data to flow into local FL nodes, enhancing responsiveness (Mishra, et al., 2023). Similarly, coupling FL with artificial intelligence for policy simulation could enable governments to test potential policies virtually before implementation, reducing risks and improving efficiency. This convergence of technologies can transform FL from a technical innovation into the backbone of global climate intelligence, underscoring its pivotal role in shaping the future of climate technology (Rane, et al., 2025).

10.7.3 Addressing Persistent Challenges

Nevertheless, the future trajectory of FL in climate action is not without significant challenges. Persistent issues of data heterogeneity necessitate harmonization efforts and the establishment of standardized protocols to ensure interoperability across diverse sources. Equally pressing is the disparity in computational infrastructure between developed and developing nations, which calls for targeted investments, cloud-enabled solutions, and international funding mechanisms to bridge this gap. Furthermore, robust governance and trust frameworks will be essential to guarantee that federated models operate in a fair, transparent, and inclusive manner. In the absence of such safeguards, there is a risk that FL could reproduce existing global inequalities, enabling data-rich and technologically advanced regions to dominate outcomes while marginalizing vulnerable communities (Raza, et al., 2025). The potential risks of not implementing FL are significant, underscoring the importance of ethical considerations, particularly equitable benefit-sharing and fairness in model outputs, which must remain central to the integration of FL within future climate frameworks (Le, et al., 2022).

10.7.4 Pathway to 2030

Integrating FL into the Climate Action Framework for 2030 will support progress towards the ambitious goals of achieving carbon neutrality, transitioning to renewable energy, and building climate-resilient infrastructure. This integration involves creating global platforms for the development and open access to federated climate models, which will be available to policymakers, industry stakeholders, and local communities. Additionally, it includes enhancing capacity in areas most vulnerable to climate change, enabling these communities to actively participate in and benefit from federated networks. If implemented effectively, FL can help narrow the gap between current emission trajectories and the 1.5°C target set by the Paris Agreement.

10.8 Proposed Framework for Federated Learning in Climate Action 2030

The proposed framework aims to make FL a key part of the 2030 action agenda for the climate change action plan. The framework consists of four layers: Data Sources, FL Core, Global Intelligence Modules, and the GFCIN. Each layer will address specific privacy, scalability, inclusivity, and decision support challenges in climate action.

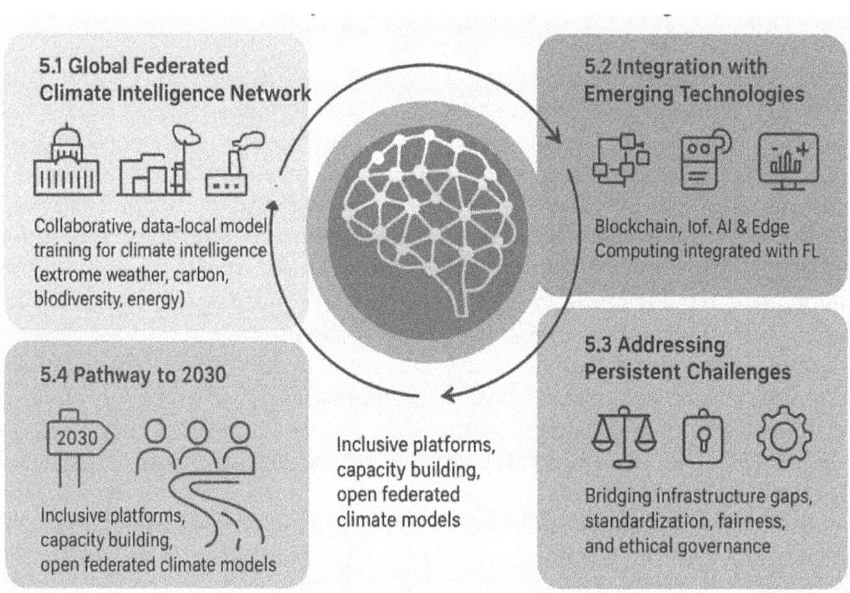

FIGURE 10.3
Future outlook of Climate Action Framework 2030 with Federated Learning, integrating collaboration, emerging technologies, and inclusive pathways for global climate intelligence.

10.8.1 Data Sources Layer

Climate-related data originates from highly diverse and distributed repositories. Sources include meteorological stations, satellites, IoT-enabled climate sensors, industrial emission records, agricultural datasets, biodiversity indices, and socio-economic resilience indicators. These datasets are inherently heterogeneous, encompassing structured databases, semi-structured logs, and unstructured reports such as remote sensing imagery. Traditionally, centralized aggregation of such data raises concerns over sovereignty, ownership, and security. In the proposed model, pre-processing operations such as normalization, encoding, and feature extraction are conducted locally to ensure data quality, while raw datasets remain within their respective domains.

10.8.2 Federated Learning layer

This sublayer emphasizes the FL process, the fundamental mechanism for collaborative model training without requiring data centralization. Each participating institution (node) trains a localized model on its respective dataset, encompassing convolutional neural networks for satellite imaging, recurrent neural networks (RNNs/LSTMs) for time-series climate forecasting, or ensemble approaches for emissions trend analysis. Instead of disseminating raw or sensitive data, the nodes convey encrypted model changes (such as weights or gradients) to a central aggregator. The emphasis is on protecting privacy and security, as these updates are protected using secure aggregation protocols and methods such as differential privacy and homomorphic encryption, guaranteeing that no individual contribution can be deconstructed (Sun, et al., 2022). The aggregator employs algorithms like Federated Averaging (FedAvg) and its sophisticated variations (e.g., FedProx, FedDyn) to consolidate local updates into a cohesive global model; after that, the global model is shared with all the local nodes. So, such an iterative process within the sublayer improves model accuracy, resilience, and international collaboration while maintaining data sovereignty (Konda, et al., 2024).

10.8.3 Global Intelligence Layer

The Global Intelligence Modules are the functional foundation of the proposed FL-enabled Climate Action Framework 2030. This layer converts scattered and heterogeneous data into actionable insights, ensuring that climate action plans are data-driven, inclusive, and adaptable to different geographic circumstances. Each module functions as a domain-specific intelligence layer but works together to provide a comprehensive climate response system.

1. **Climate Prediction and Modelling:** Federated integration of regional climate datasets improves high-resolution forecasting of extreme events such as floods, droughts, cyclones, and heatwaves.
2. **Carbon Monitoring:** Cross-industry FL enables verifiable carbon accounting and emission tracking without exposing commercially sensitive or politically sensitive data (Konda, et al., 2024).
3. **Energy Optimization:** Federated reinforcement learning supports renewable energy integration, predictive load balancing, and grid optimization while safeguarding operational privacy.
4. **Agriculture and Food Security:** FL-driven models predict crop yields, optimize irrigation/fertilizer use, and enable early detection of pests or diseases by leveraging local datasets across geographies (Cowlishaw, et al., 2025).
5. **Biodiversity and Ecosystem Monitoring:** Shared federated models allow transnational monitoring of species migration, extinction risks, and ecosystem health, supporting global conservation strategies (Vamshikrishna, et al., 2024).

10.8.4 Climate Action Network Layer

The final layer operationalizes a GFCIN that establishes connections between governments, industries, research institutions, and communities. At this layer, blockchain can be implemented, so blockchain-based audit trails guarantee accountability and equitable recognition of contributions by providing transparent and tamper-proof recordings of participation (Yuan, et al., 2024). Similarly, a cloud-native federated infrastructures ensure scalability and participation from both technologically advanced and resource-constrained regions at this layer. So, at this layer, the smart contracts facilitate incentive mechanisms for benefit-sharing, and standardized protocols with APIs facilitate seamless integration across borders to attain interoperability. Moreover, the intelligence generated is incorporated into global decision-support interfaces, which subsequently enhances evidence-based policymaking, Nationally Determined Contributions (NDCs), and international negotiations under the Paris Agreement (Godavarthi, et al., 2025).

(Figure 10.4) emphasizes the layered architecture of FL for climate action, illustrating the workflow of the proposed framework. At the fundamental level, distributed datasets from meteorological, industrial, agricultural, and ecological sources are pre-processed and locally owned. After that, in the FL layer, a computationally secure privacy-preserving method offers secure aggregation of local models of various local nodes at the data source layer. In this manner, the global models are specialized into application modules, which include predictive climate modelling, carbon accounting, renewable energy optimization, agricultural resilience, and biodiversity monitoring. The GFCIN is a governance and trust-enabled ecosystem that ultimately converges the outputs. Thus, the proposed framework can utilize blockchain-based audit trails and interoperability standards to deliver transparent, scalable, and equitable intelligence to policymakers, industries, and communities worldwide.

10.9 Conclusion

Climate change demands immediate, cooperative, and evidence-driven solutions that transcend political will and traditional frameworks. This chapter has illustrated how FL could be a revolutionary approach to tackle issues such as data fragmentation, sovereignty concerns, and unequal access to technology. FL facilitates decentralized yet collaborative intelligence, enabling stakeholders to collectively develop predictive climate models, enhance renewable energy systems, monitor emissions, preserve biodiversity, and shape evidence-based policies, all while retaining control over sensitive data. Simultaneously, challenges like data heterogeneity, infrastructure gaps,

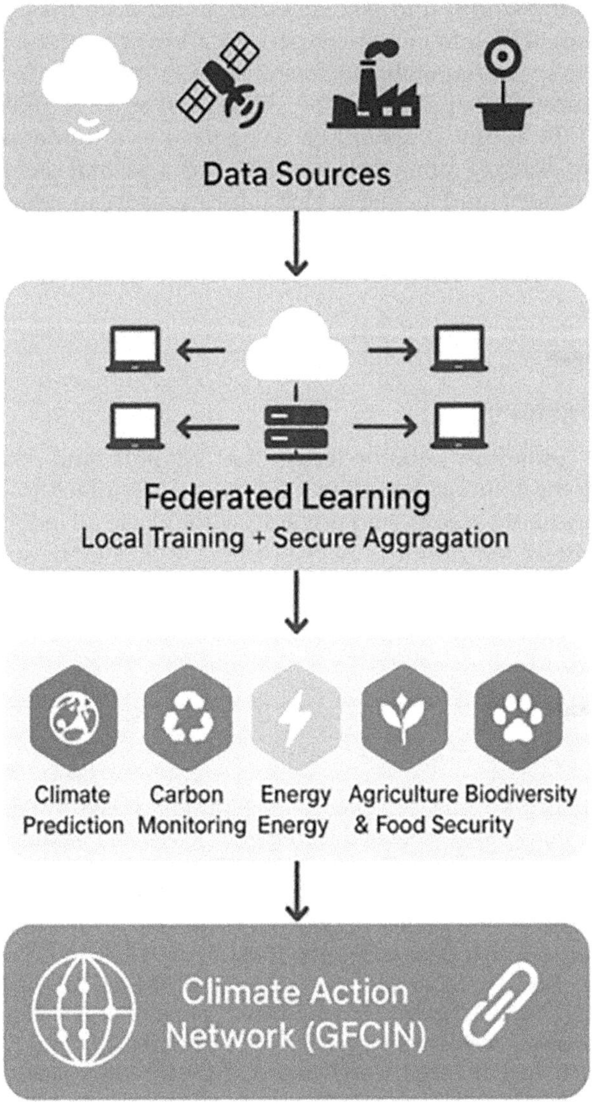

FIGURE 10.4
Proposed framework for Federated Learning in Climate Change Action Plan – 2030.

governance frameworks, and ethical issues underscore the critical need to carefully integrate FL into global systems. By addressing these issues, FL has the potential to ensure equitable inclusion and fair treatment for all, instilling hope for its successful application. FL should not be viewed as just another innovation in the future; it should be recognized as a fundamental element of the Climate Action Framework for 2030. It is a pivotal tool for shaping a sustainable, resilient, and inclusive global future, as it can safeguard privacy, promote fairness, and enhance intelligence across borders.

Acknowledgement

The authors gratefully acknowledge the support and encouragement received from the National Institute of Advanced Manufacturing Technology (NIAMT), Ranchi. We also extend our sincere thanks to all individuals whose direct and indirect contributions have been invaluable in the successful completion of this study.

References

Ahmed, L., Ahmad, K., Said, N., Qolomany, B., Qadir, J. and Al-Fuqaha, A., 2020. Active learning based federated learning for waste and natural disaster image classification. *IEEE Access*, 8, pp.208518–208531.

Cowlishaw, R., Longépé, N. and Riccardi, A., 2025. Balancing centralisation and decentralisation in federated learning for Earth Observation-based agricultural predictions. *Scientific Reports*, 15(1), p.10454.

Godavarthi, D., Jose, D., Mohanty, S.N., Medani, M., Kallel, M., Abdullaev, S. and Khan, M.I., 2025. Federated learning-based semantic segmentation framework for sustainable development. *Egyptian Informatics Journal*, 30, p.100702.

Killeen, P., Lin, C., Li, F., Kiringa, I. and Yeap, T., 2025. IoT-based smart farming architecture using federated learning: a nitrous oxide emission prediction use case. *ACM Journal on Computing and Sustainable Societies*, 3(2), pp.1–38.

Konda, B.R., Tummala, V.M.R., Ganugapenta, S.K.R., Siraparapu, P.K., Hazra, A. and Gurusamy, M., 2024. October. Fedwavg: mitigating model contamination in uav networks through federated weighted average for weather forecasting. In *2024 IEEE 100th Vehicular Technology Conference (VTC2024-Fall)* (pp. 1–5). IEEE.

Kulovesi, K. and Oberthür, S., 2020. Assessing the EU's 2030 climate and energy policy framework: Incremental change toward radical transformation?. *Review of European, Comparative & International Environmental Law*, 29(2), pp.151–166.

Le, D.D., Tran, A.K., Dao, M.S., Nguyen-Ly, K.C., Le, H.S., Nguyen-Thi, X.D., Pham, T.Q., Nguyen, V.L. and Nguyen-Thi, B.Y., 2022. Insights into multi-model federated learning: An advanced approach for air quality index forecasting. *Algorithms*, 15(11), p.434.

Maurya, S.P., Yadav, A.K., & Singh, R. (Eds.), 2022. *Modeling and Simulation of Environmental Systems: A Computation Approach* (1st ed.). CRC Press. https://doi.org/10.1201/9781003203445

Mishra, R., Ramesh, D., Bellavista, P. and Edla, D.R., 2023. Redactable blockchain-assisted secure data aggregation scheme for fog-enabled internet-of-farming-things. *IEEE Transactions on Network and Service Management*, 20(4), pp.4652–4667.

Mrabet, M. and Sliti, M., 2025. *Towards Secure, Trustworthy and Sustainable Edge Computing for Smart Cities: Innovative Strategies and Future Prospects*. IEEE Access.

Nanda Assobjio, B.Y., 2024. *Towards Using Federated Learning to Improve Generalization in Weather Forecasting* (Doctoral dissertation, Polytechnique Montréal).

Nooh, S.A., 2025. *Optimizing Low Carbon Sustainable Environmental Monitoring with Consumer Technology: An Iot-Driven Federated Learning Approach for Edge Computing Optimization*. IEEE Transactions on Consumer Electronics.

Pimenow, S., Pimenowa, O., Prus, P. and Niklas, A., 2025. The impact of artificial intelligence on the sustainability of regional ecosystems: Current challenges and future prospects. *Sustainability*, 17(11), p.4795.

Qiu, X., Parcollet, T., Fernandez-Marques, J., Gusmao, P.P., Gao, Y., Beutel, D.J., Topal, T., Mathur, A. and Lane, N.D., 2023. A first look into the carbon footprint of federated learning. *Journal of Machine Learning Research*, 24(129), pp.1–23.

Rane, N., Mallick, S.K. and Rane, J., 2025. Artificial intelligence-driven climate change adaptation and ecosystem resilience. *Available at SSRN 5362147*.

Raza, M.O., Umar, A., Rasheed, J., Asuroglu, T. and Alsubai, S., 2025. Federated transfer learning for distributed drought stage prediction. *Discover Artificial Intelligence*, 5(1), p.55.

Scorza, F. and Santopietro, L., 2024. A systemic perspective for the sustainable energy and climate action plan (SECAP). *European Planning Studies*, 32(2), pp.281–301.

Soergel, B., Kriegler, E., Weindl, I., Rauner, S., Dirnaichner, A., Ruhe, C., Hofmann, M., Bauer, N., Bertram, C., Bodirsky, B.L. and Leimbach, M., 2021. A sustainable development pathway for climate action within the UN 2030 Agenda. *Nature Climate Change*, 11(8), pp.656–664.

Sun, T., Li, D. and Wang, B., 2022. Decentralized federated averaging. *IEEE Transactions on Pattern Analysis and Machine Intelligence*, 45(4), pp.4289–4301.

Thakur, D., Guzzo, A., Fortino, G. and Piccialli, F., 2025. Green federated learning: A new era of green aware AI. *ACM Computing Surveys*, 57(8), pp.1–36.

Vamshikrishna, A., Ramesh, D., Mishra, R. and Mohammad, N., 2024. Sustainable Healthcare 5.0: Integration of IoT and blockchain technology with federated learning model for securing healthcare data. In Misra, S., Siakas, K., Lampropoulos, G. (eds) *Artificial Intelligence of Things for Achieving Sustainable Development Goals* (pp. 161–180). Springer Nature Switzerland.

Yuan, W., Ye, G., Zhao, X., Nguyen, Q.V.H., Cao, Y. and Yin, H., 2024. Tackling data heterogeneity in federated time series forecasting. *arXiv preprint arXiv:2411.15716*.

Zuccaro, G., Leone, M.F. and Martucci, C., 2020. Future research and innovation priorities in the field of natural hazards, disaster risk reduction, disaster risk management and climate change adaptation: A shared vision from the ESPREsSO project. *International Journal of Disaster Risk Reduction*, 51, p.101783.

Index

For Product Safety Concerns and Information please contact our EU
representative GPSR@taylorandfrancis.com
Taylor & Francis Verlag GmbH, Kaufingerstraße 24, 80331 München, Germany

www.ingramcontent.com/pod-product-compliance
Ingram Content Group UK Ltd.
Pitfield, Milton Keynes, MK11 3LW, UK
UKHW020202200926
475702UK00004B/572
* 9 7 8 1 0 3 2 8 6 6 4 7 5 *